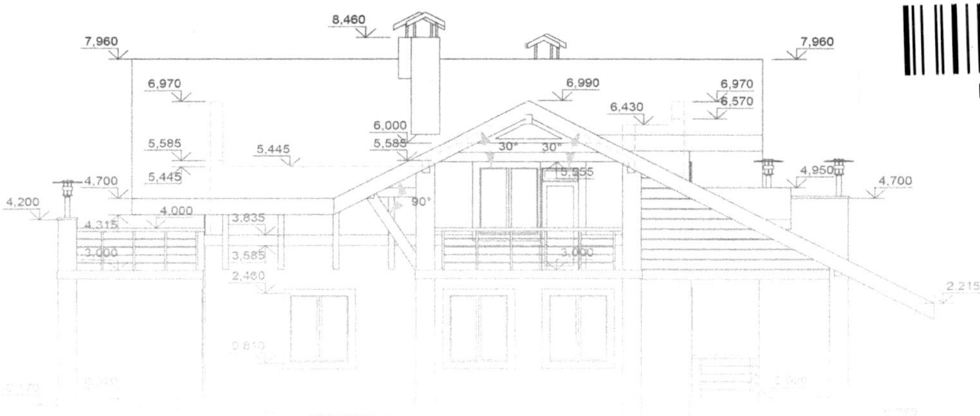

THE PRODUCTION HOMEBUILDER

FIRST EDITION

Will Holder

Texas State University—San Marcos

cognella
SAN DIEGO

Bassim Hamadeh, CEO and Publisher
John Remington, Executive Editor
Carrie Baarns, Manager, Revisions and Author Care
Gem Rabanera, Senior Project Editor
Susana Christie, Senior Developmental Editor
Jordan Krikorian, Editorial Assistant
Celeste Paed, Associate Production Editor
Jess Estrella, Senior Graphic Designer
Kylie Bartolome, Licensing Coordinator
Stephanie Adams, Senior Marketing Program Manager
Natalie Piccotti, Director of Marketing
Kassie Graves, Senior Vice President, Editorial
Jamie Giganti, Director of Academic Publishing

Printed in the United States of America.

cognella® | ACADEMIC PUBLISHING
3970 Sorrento Valley Blvd., Ste. 500, San Diego, CA 92121

Table of Contents

ACTIVE LEARNING

This book has interactive activities available to complement your reading.

Your instructor may have customized the selection of activities available for your unique course. Please check with your professor to verify whether your class will access this content through the Cognella Active Learning portal (http://active.cognella.com) or through your home learning management system.

cognella® | ACTIVE LEARNING content

ACCESSING WEB-BASED RESOURCES

This book has QR codes available to complement your reading.

The author has selected additional web-based content for further engagement using QR codes, which are intended for those who have purchased print copies of the book. Those who have purchased a digital copy may simply click on the corresponding hyperlinks. Please check with your professor to confirm whether your class will access this content independently or collectively.

Cognella maintains no responsibility for the content nor availability of third-party links. However, Cognella makes every effort to keep texts current. Broken links may be reported to studentreviews@cognella.com. Please include the book's title, author, and 7-digit SKU reference number (found below the barcode on the back cover of the book) in the body of your message.

cognella
SAN DIEGO

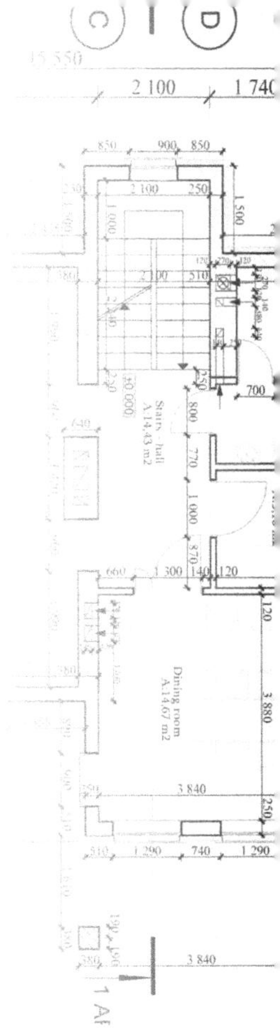

Preface

Homebuilding gave this professor the profession of his dreams. After a more-than-thirty-seven-year career in the homebuilding industry, this homebuilder-turned-professor has been given another homebuilding career teaching, inspiring, and connecting construction students to the life-changing careers awaiting them in the homebuilding industry. This professor of practice was not the most qualified homebuilder to write this textbook but only the one who found the Oncor opportunity give back to the industry I love. I am twice blessed by the homebuilding industry!

Thank you to David Weekley Homes for giving this builder-professor his start in homebuilding. This professor responded to an ad in the *Houston Chronicle* titled "We'll teach you to build houses!" by Weekley Homes, and that started it all. Thank you David Weekley Homes on behalf of the thousands of others like myself whose homebuilding careers started at that great company. Additional thanks go to the homebuilding professionals at David Weekley Homes who assisted this professor in the early stages of this curriculum development. Teaching people to start and build a great homebuilding career is nothing new to that organization as their grateful alumni are found in leadership positions throughout the homebuilding industry.

Thank you to John Walsh, an accomplished residential land developer and homebuilder, who perhaps found his greater calling as a real estate educator who founded and built the University of Houston's C.T. Bauer Graduate Real Estate Certificate Program. Thanks to John for giving this professor his first university teaching opportunity at the University of Houston. That opportunity ignited my teaching passion and opened my eyes to the need for new residential college curriculum. I first envisioned writing this textbook while teaching at University of Houston.

Thank you to Dr. Gary Winek for hiring and mentoring this professor of practice as a new construction science and management professor and for your valuable input on this curriculum. However, Dr. Winek's lasting impact on our industry will be his decades of personal passion for his students and to the building science. Dr. Winek's commitment to construction education built and nurtured from inception the great Construction Science and Management program at Texas State University . What a marvelous legacy!

This professor is especially thankful for his twenty-five years at Trendmaker Homes, where much of what is included in this textbook was learned from an amazing team of professionals representing every homebuilding discipline. A personal thanks goes to Joe Mandola, a true master production homebuilder, for his thirty years of support and professional partnership. Working with all my former co-workers at Trendmaker Homes was a professional honor, and I always look back fondly at the great years we enjoyed together.

Thank you to Kerry Gilbert and Michelle Stoops for their contribution of the "Land Planning: Vision and Design Process" chapter. Their well-earned reputation and prolific body of land-planning work made them the ideal authors for this important topic. Residential land development deserves its own textbook and entire college course. Kerry's and Michelle's contribution is an excellent step in that direction.

Thank you to the Texas Association of Builders for their support of homebuilding college curriculum by contributing their important chapter, titled "The Homebuilders Association," aimed at introducing future homebuilding company owners and executives to the builders associations early in their careers. It is important that we reach future association members early!

Thank you to Texas State University for supporting residential construction education and having the vision to create a professor of practice, residential construction, position to help graduates realize the great opportunities awaiting them as future homebuilders. Texas State University is an innovator in construction education with professors of practice in residential and commercial construction, both instructors being former company presidents in their respective sectors. The Texas State Department of Engineering Technology's innovative approach to construction education brings industry-tenured instruction into the classroom.

Will Holder, editor/author

Professor of Practice, Residential Construction;

former homebuilder; Trendmaker Homes;

2012 president, Greater Houston Builders Association

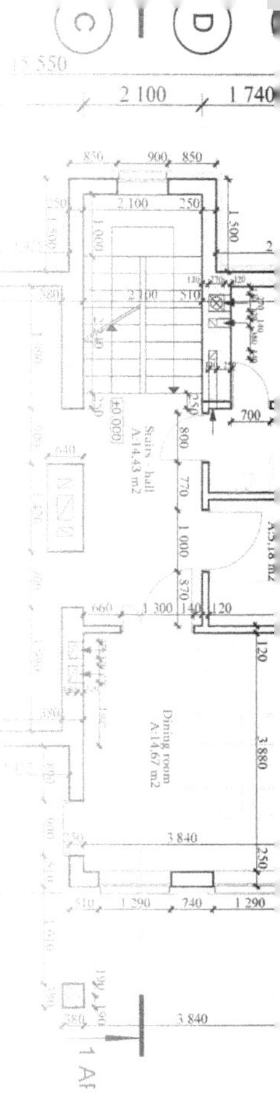

Introduction

This textbook and its course companion components were written to introduce and promote the homebuilding industry to construction college students and to prepare those graduates who choose to become homebuilders for a fast career start and an exciting and rewarding career journey.

Homebuilding is more than a career; it is an honorable profession. Those who pursue a life in homebuilding will someday proudly look back at the beautiful homes and communities they built and the homeowners' lives they improved. Homebuilding is a retail industry; we create our new home products and sell them face to face directly to the end users, the happy homeowners.

The Production Builder, and its eight companion photo tours through the thirteen homebuilding stages, is university curriculum designed to introduce construction students to the homebuilding industry and to the residential stages and sequence of construction. Teaching construction students the stages and sequence of the new home construction process is a great foundation to start their residential career with confidence and the fundamental knowledge that will allow new construction graduates to be contributors from day one in the industry.

The Production Builder course is designed to provide all university construction science and management programs a means to teach residential construction the same way builders have historically taught new builder employees—by learning the stages and sequence of construction from a construction manual. Using a manual to instruct new builder employees in the home construction process follows a time-tested teaching method that is now being brought into the college classroom.

Homebuilding companies are only beginning to realize the great fountain of new talent being trained in the American Council for Construction Education (ACCE)–accredited universities across North America. There are ACCE-accredited two-year and four-year construction programs in every major housing market in America, and those programs are preparing tomorrow's homebuilders to be ready to start their careers as entry-level managers and contribute immediately upon graduation. With homebuilding specific curriculum, ACCE-accredited universities can now become even greater sources of enthusiastic, well-educated, and diverse new homebuilding talent for our industry.

ACCE
AMERICAN COUNCIL FOR CONSTRUCTION EDUCATION
PARTNERSHIP FOR EXCELLENCE

Image 0.1

Homebuilders seeking to connect with one of the more than ninety ACCE-accredited university construction programs should visit www.acce-hq.org for more information and a list of accredited universities.

WATCH AT https://tinyurl.com/2p8ntw2d

This textbook and manual and its eight companion photo presentations are a work in process. All industry input is welcome. Would you like to contribute to the next edition of *The Production Homebuilder*? Feel free to reach out to me directly at fwh10@txstate.edu with any thoughts, edits, or additions.

This textbook is just the starting point for a new homebuilding college curriculum!

Figure Credit

IMG 0.1: Copyright © by American Council for Construction Education (ACCE). Reprinted with permission.

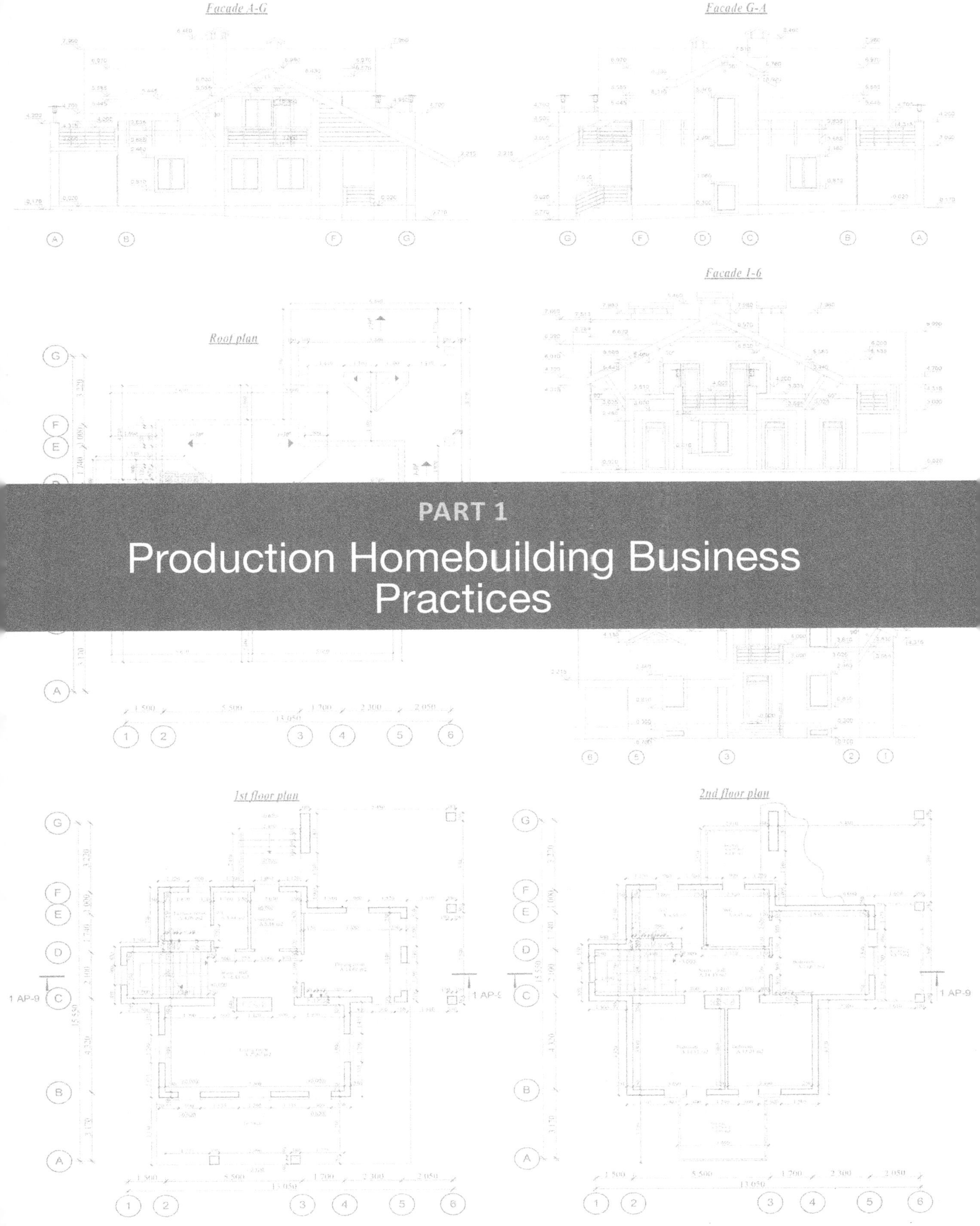

PART 1
Production Homebuilding Business Practices

RESIDENTIAL INDUSTRY PROFILE

CHAPTER 1

INTRODUCTION

Construction Science and Management (CSM) students are preparing for great careers in any one of the four major construction sectors. The residential sector is uniquely connected to other construction sectors because wherever you find active new home construction, you will also find active construction of new schools, churches, and retail and commercial centers needed to service the new residents. Residential construction is America's leading economic indicator because the new "rooftops" we build lead the way for the commercial, heavy civil, and industrial construction sectors. Residential construction is the largest of the four construction sectors, representing as much as 5 percent of America's gross domestic product.

Because residential construction always precedes the other construction sectors (commercial, industrial, and heavy civil) and is the goliath of construction sectors, it is important for all CSM students to have professional knowledge and an understanding of the residential industry, no matter which construction sector they may ultimately find their career. Understanding the drivers in the construction industry at large—seeing the big picture—is a valuable perspective for all future construction industry leaders.

Homebuilding Is a Retail Service Industry

CSM students should view the single-family residential industry as a retail service industry rather than a construction industry. Every new home is sold face-to-face by the building company, one at a time, directly to the end user, the homeowner. That is the purest definition of retail. Due to the intimacy of the home product and level of customer interaction it takes to build someone a new home, the most successful builders evolved from being construction companies into construction service providers. Residential construction is a retail service.

The retail component of homebuilding is an important point to be understood by CSM students because homebuilding requires the constructor to satisfy an emotional, nontechnical customer. To become a homebuilder, one must master not only the actual construction process of the home but also the skills needed for a face-to-face relationship with the customer.

Homebuilders collaborate with clients through what can be a stressful process. Today's homebuilders must be competent constructors, creative space designers, brand marketers, accomplished salespeople, and smooth retail store operators that connect with and manage happy buyer clients through a long and complex retail relationship.

For a family to uproot from an existing home, buy, and move into a new home, the process normally involves a significant life event, such as a new baby. In addition, the home is the single biggest retail purchase for most families, which carries significant legal implications and duties to be performed by the builder.

Builders have evolved into employees within a service-based business, not just home product companies. CSM students choosing to pursue their career in homebuilding will need to rely on their retail people skills as much as their technical construction skills.

What Is *Residential?*

Residential is a big word that needs some definition. Starting with the broadest definition, the residential industry designs and build complete communities and the dwellings intended for full-time living.

Residential homes include:

- Apartment homes
- Townhomes
- Duplex homes
- High-rise homes
- Patio/garden homes
- Single-family homes

The common denominator of all the above home types is this is where people will live full-time. The residential industry designs and builds homes for year-round living to meet the year-round needs and desires of all life stages and family sizes.

Homes may be owned or rented, but the distinction of year-round residency is the key. This is because people are extremely specific about where they want to live full-time versus where they may spend a week's vacation. The residential industry does not include hotel or vacation rental communities or other forms of temporary housing. Although the construction process is similar and related, the industry that develops these types of properties work off a completely different set of business and financial metrics. The for-rent-for-the-weekend business model is completely different from the buy-and-live-here model.

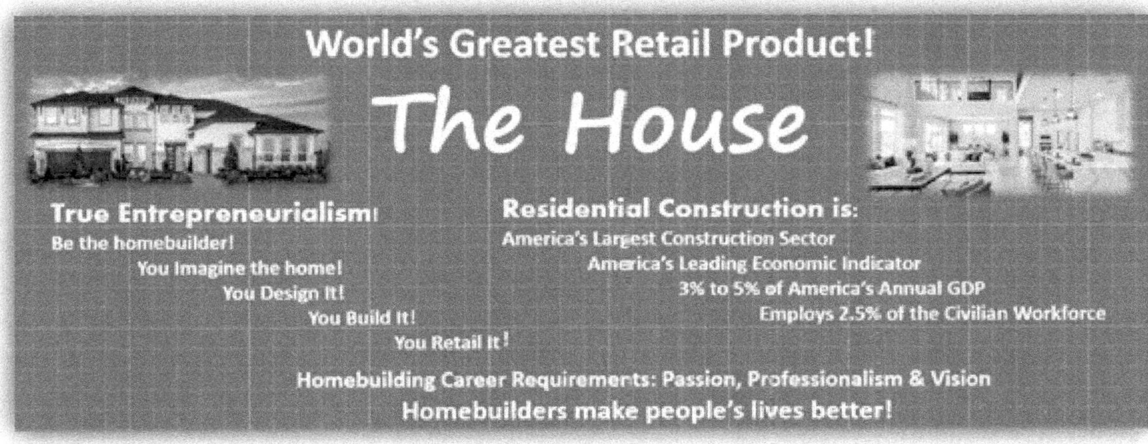

FIGURE 1.1 The House, above, is a display case backdrop at Texas State University located in the Roy F. Mitte building, where CSM students have their classes. The display case exhibits a large collection of homebuilding books, magazines, and artifacts dating back to the mid-1300s and extending into the 2000s.

Multifamily and Single-Family

Construction students will notice that the residential construction industry is further subdivided into single-family and multifamily housing. This can be confusing because some news articles write about *single-family starts*, whereas another article discusses *housing starts*, which includes both single-family and multifamily starts. *Housing starts* is the broadest definition, counting a combination of all multifamily and single family starts. Any report on housing starts may also speak to *single-family starts*, which is specific to detached, single-family dwelling units or homes. Students should look for and understand this distinction in any report or news article.

CSM students should note that a career in multifamily construction is a different career from one in single-family construction. This professor spent six years in multifamily at one point and returned to single-family to change my work-life balance.

A career in multifamily construction will share many of the same characteristics as those of a career in commercial construction. This is because the multifamily project is a large land development and building construction project with a defined beginning and end that builds at a pace disconnected to the project's rate of sales or rentals. The construction goal is to start, build, complete, and exit the entire project as soon as possible.

Single-family construction, on the other hand, moves at a pace directly connected to the pace of sales. Single-family homes are started individually or in small groups called *panels*, normally a group of about five contiguous or nearby lots that are started and built together in a panel group. The pace of new home starts will increase or decrease to match the pace of home sales. All single-family home construction is tied to selling homes.

From 1950 to 2000, the United States built an average of around one million single-family homes per year. In America today, there are more than 1.5 million *total housing starts* annually, of which single-family starts make up more than one million.

The Next Challenges

Home design can be simple or complex. Homes are changing all the time based on buyers' changing tastes. Homes are fashion and have a style. The ways in which homes are built also changes. At the beginning of this professor's career, homes were designed with individual rooms separated by walls for every function, including dining, cooking, and family time. Today, it is common to find open floor plans, where the kitchen and living and dining rooms are all in one big open room. It's in fashion, and that fashion change drives new sales to fashion-sensitive buyers.

Today's CSM students need to see beyond the styles and the methods of building homes used today. The form of tomorrow's homes is up to today's CSM students to envision and create. The only thing we know for sure is that people will always need a home. Homes are essential! There are two challenges CSM students will face in their future career: the methods by which we will build homes in the future and the the styles or characteristics that will be desired and demanded in tomorrow's homes.

THE EARLY RESIDENTIAL PROFESSION

FIGURE 1.2 This 1888 publication was one of the earliest American homebuilding industry publications offering advertising opportunities for material vendors of all kinds and articles educating builders about new technologies such as private line telephones, a "hot water apparatus," and lead paint. Builders were always looking for the next new product!

Throughout the millennium, the craft of building homes has been a highly marketable skill. It is also a profession that carries a profound professional responsibility. Professionalism in home building has early roots dating back to a Babylonian residential building code written in stone circa 1780 BC, the Code of Hammurabi. That homebuilding code stated that if a builder built a home and it collapsed and killed a family member of the homeowner, then that builder or some family member of the builder should also be put to death. Now, that is a stiff building code indeed! Fortunately, we have new building codes today.

CSM students considering homebuilding as a career are entering a proud and essential profession of providing the service of home construction for their customers and a new neighborhood for the overall community.

In early America, the European immigrants built timber homes. Joiners were the early craftsmen, building sturdy wooden home frames by joining 8-feet-by-8-feet or 6-feet-by-12-feet hand-honed timbers. Light frame construction came into existence only in the 1840s, brought on due to the availability of plentiful factory-made nails and water-powered sawmills that could quickly cut smaller dimensional lumber, or boards, from logs and timber. Those boards and store-bought nails quickly gained popularity and were fashioned into a much-needed construction method that continues through today and is known as *light frame construction*.

FIGURE 1.3 This timber frame house was built in San Felipe, Texas, in the 1820s and is representative of the timber homes being built in Texas prior to the introduction of dimensional 2X boards.

In the mid-1800s, learning how to build using boards and lumber and attach them with nails to build a strong and durable house frame was an evolutionary process of trial and error. Boards and nails were easily carried in a wagon, and only two people working together were able to construct the frame. America needed a lot of new homes as the country's population grew, and from that wave of rising demand, the industry began to adopt, learn, and perfect the wood frame house.

The Evolving Profession

American's first homebuilders evolved from the local, skilled wood joiners who served their own town building homes one at a time. These homebuilders' productivity was limited to their own knowledge and own two hands. Architects were the next logical professionals who knew how to design and build a home and came to offer that service to their communities. Over time, the educated architects who worked with these skilled craftspeople found the latter could organize into skilled groups of carpenters, masons, and roofers to increase productivity on the architects' new home projects while acting on behalf of the architects and directing other workers.

Those architects came to understand that they could not draw the house plans and supervise construction at the same time. So, they began to hire competent construction assistants from their skilled crews to oversee the daily construction activities on their job sites. Those assistants, over time, became competent homebuilders under their own name. These were builders with solid technical skills and construction knowledge that often exceeded even that of the architect builders.

The most competent of those on-site constructors then began to build homes under their own name and attract business based on their reputation. These constructors became the residential contractors. This new breed of general contractor homebuilders were not only skilled and knowledgeable constructors; they were also becoming the risk takers who built and sold their new homes in the open market. And from those contractor-builders, the modern homebuilding industry was born.

Real Estate Became Very Big Business, and Why!

What was once a local cottage industry changed forever in the 1950s, when there were approximately two million homes built in a single year in the United States. Even by today's standards, that was a gigantic year for housing starts, but back then it was the birth of a new age in the residential industry. Today's version of the high-volume, high-production homebuilder had their beginning in the 1950s.

New home production was achieving previously unimaginable new levels of sales and construction velocity. Most notable was Levittown, Pennsylvania, where as many as 30 homes a day were completed during the community's peak years in the 1950s.

Conventional wisdom says that the 1950s boom was ignited by the family formation at the end of World War II (WWII). Although it is true that family formation and the demand for new homes was roaring after WWII, the biggest game changer for the residential industry was the proliferation of new forms of construction lending and the broader availability of home mortgages. All kinds of new financing for land developers, homebuilders, and homebuyers sparked the mid-century roar to homeownership that continues today. Mortgages and construction and development lending unlocked this giant industry.

Land and Homes as Collateral at Every Step

Real estate development and homebuilding both have one unique financing characteristic: your land and its improvements, the normal inventory of a developer or homebuilder, can be used as collateral to finance the development or home project.

In the case of developers, their raw material, land, has its own intrinsic value that can be borrowed against. Hence, banks began loaning developers a percentage of the total capital required to build a new residential subdivision. Land developers found that this financial leverage made it financially feasible to develop the high volume of new lots to match the high housing market demand.

Next, builders found that they could also borrow against the value of a lot with a partially built home on it, and that financial leverage allowed builders to buy more lots and build more homes. Construction lending multiplied the builders' working capital and thereby multiplied their ability to build more homes and enter new communities and city markets.

The next link in the residential chain are the want-to-be homeowners. Mortgages also allowed buyers to borrow 80 percent, or more, of the value of the completed home. The retail mortgage loan was the terminal financing needed to secure each preceding step in the financial system that ultimately delivered Americans the new home they wanted and needed. Construction students must understand that no matter how much buyers want to buy new homes, it is the availability of retail mortgages that make possible the homebuilding industry that we know today. In conclusion, the land development and home construction lending and the mortgage industry together form the hinge pin that holds together the entire homebuilding industry.

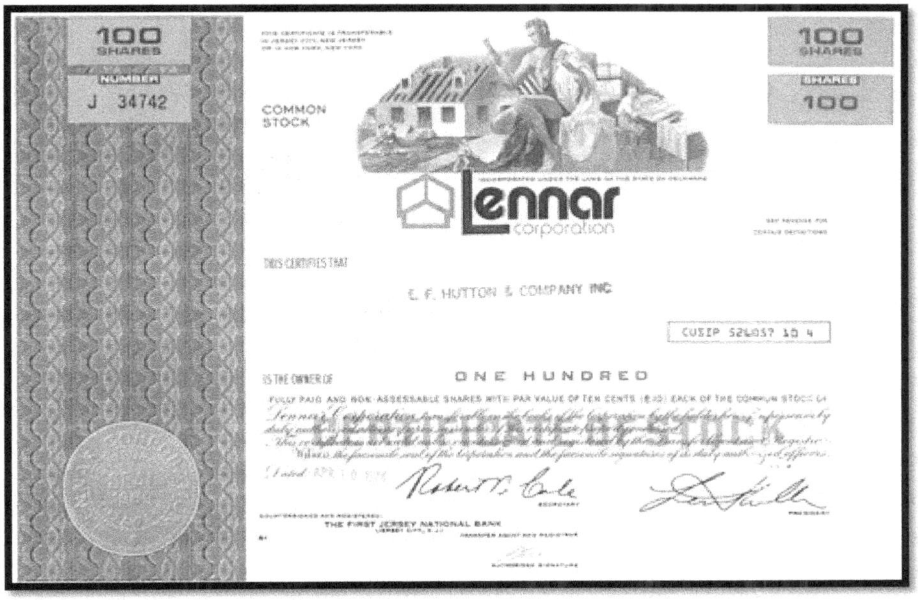

FIGURE 1.4 Access to Wall Street money fueled homebuilding to new growth and sophistication. The financial leverage available by investing in residential land development and homebuilding did not go unnoticed by Wall Street, as seen in this 1974 Lennar Corporation stock certificate. Notice that the original date of incorporation was 1969.

FIGURE 1.5 The idea that owning a home was an investment and not an expense was new to Americans in the 1950s. Mortgages made owning a home possible, so the entire industry went about spreading the good news that, in the long run, owning is less expensive than renting. The acceptance of the home as an investment was then—and still is today—the foundation on which our industry is built.

PUBLIC AND PRIVATE BUILDERS

Homebuilding is a gigantic industry that builds one million to two million homes a year, with tens of thousands of building companies participating. That huge collection of builders may be broken into two groups. The first group are those homebuilding companies with public ownership, and the second are those with private ownership.

Public builders are those whose common ownership shares trade publicly on Wall Street, and the public companies number only about 18 to 20 companies, based on the latest mergers. Although there are not many *public builders*, they build an impressive more than 25 percent of all homes built each year. These industry giants are market-tried and market-tested companies indeed, with each public company is building thousands or tens of thousands of new homes every year.

These *public companies* are great places for CSM students to start and build their residential career. Public companies are big employers and are almost always actively hiring in cities all around the country. According to the 2021 Builder 100 List data, the publicly owned builders built approximately 300,000 homes per year, for a dollar volume of over $130 billion! These public companies are very large as measured by both total revenues and total number of homes built.

Homebuilding companies with *private ownership* command an amazing 75 percent market share. Compare that to the beer industry, in which private brewers have only about a 15 percent share of the beer market. Homebuilding has always been an entrepreneurial industry in which competent builders have been able to enjoy successfully building homes under their own brand. It may be easy to start a homebuilding company, but it is not as easy to maintain and grow a homebuilding company over the decades.

Who are the biggest players in the industry today?
WATCH AT https://tinyurl.com/bdcvh4bc
2021 Builder 100 List/*Builder* magazin

IMPORTANT INDUSTRY PARTICIPANTS

Builders and developers rely on a vast array of direct and indirect industry participants who perform or directly support the actual construction and delivery of the home to the new homeowners. These collective teams of industry specialist make up this immense residential industry that is building beautiful new homes and communities across America.

Construction Partners

This group of industry participates may be placed into two related groups. First are the suppliers, also called *vendors*, and the second are the *subcontractors*. For the purposes of today's discussion, the significant distinction between a *vendor* and a *turnkey subcontractor* is whether the suppling company performs the installation or not. Does the construction partner deliver materials only, or do they deliver and install their materials?

If the partner only delivers materials, then they are a *vendor/supplier.* If the partner delivers and installs or installs the builder's materials, then they are considered a *subcontractor.*

Suppliers

The home is a magnificent assembly of premanufactured building materials and mechanical and electrical components. Those materials and components move through a highly complex supply chain as they make their way from a factory or mill to each individual job site. There is an enormous business sector of mostly privately owned distributors who stock and provide just-in-time inventory of all kinds of building materials to homebuilders. This materials distribution sector of our industry is gigantic and vital.

Over the past two decades, homebuilders have increasingly paid more attention to supply chain management. This is true in many industries but is an especially important field in homebuilding because the entire home is built from sourced and supplied materials.

Homebuilders write their annual plan and can accurately project their construction material quantity requirements over the next 12 to 24 months. With that information in hand, the purchasing professionals inside the homebuilding company must arrange and negotiate a reliable source for every component in the home. There will be more on this topic when we discuss the purchasing department.

FIGURE 1.6 Building material suppliers deliver their material for use by subcontractors.

Subcontractors

Teams of specialty subcontractors perform the physical assembly of the home. Those skilled craftspeople include licensed mechanical, electrical, and plumbing (MEP) contractors such as heating, venting, and air conditioning (HVAC) and electricians and plumbers who provide turnkey installation of the MEP systems in the homes. Subcontractors also include the many unlicensed but highly skilled craftspeople, such as carpenters, masons, tractor operators, and too many others to list here. Subcontractors are the skilled installers of everything in the home. Most new homes will require 20 to 30 subcontractors to be working on them.

Supply chain management does not overlook the subcontractors. If your plan says that you will be building 15 percent more homes next year, then you need to know if your subcontractor base has 15 percent more capacity or whether you need to source new or additional subcontractors. Planning and communication with your subcontractor partners are the keys to achieving your company's business plan in future years.

FIGURE 1.7 Subcontractors install materials bought by the homebuilder, or the subcontractors turnkey work and install materials they supplied.

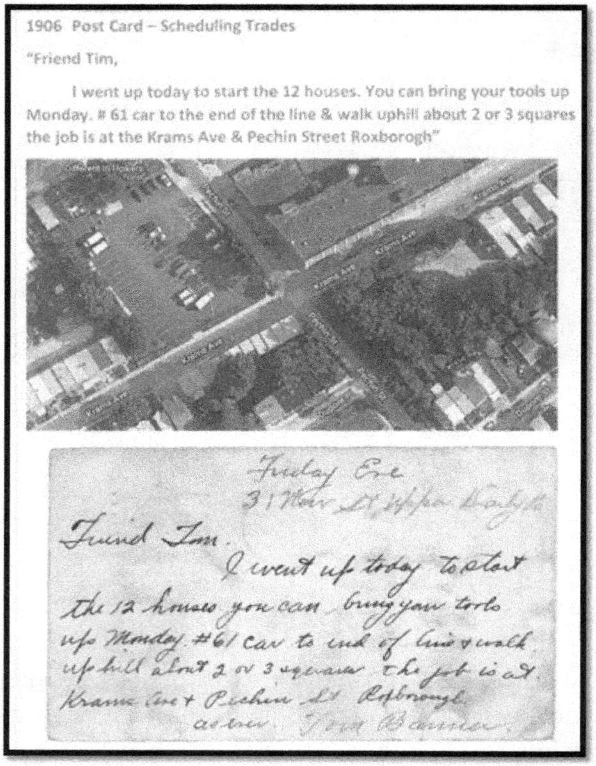

FIGURE 1.8 The skilled trades have always been the builders' primary partner in building their homes. This stamped and mailed postcard from builder Tom is reaching out to his friend Tim to bring his tools on Monday, and just a few blocks from Car 61's last stop, he will find 12 houses started and waiting for them to get to work. Today we send emails or text messages to our trade partners, but you can see that things have not changed much since 1906.

Despite decades of looking for better, faster, and cheaper ways to build homes, the on-site construction method remains the dominate method of new home construction. If this remains the case, the subcontractor will remain the production builder's most visible and valued construction partner. CSM students need their best interpersonal skills to manage this diverse and independent workforce because these homes are handmade by skilled craftspeople who work together as a well-oiled home construction machine.

Land Development and Construction Lenders

The residential industry is built upon a series of leveraged steps required in the home delivery process. Acquisition and development loans make the new community and lot possible; home construction loans make the home construction possible, and mortgages provide the permanent financing to the buyer.

CSM students will most likely have their first encounter with construction lenders when a lender representative performs a site inspection of their homes to approve a loan draw application made by their company. As future builders, CSM students should understand the importance of the lender and their routine site inspection, be ready to assist or answer questions, and always allow good access to the home by the lender representative.

The Mortgage Company

Financing that final step, homeownership, in the new home system are the mortgage companies. Because most homebuyers cannot afford to pay 100 percent cash for their new home, they need to secure a home loan or mortgage. The mortgage company requires a minimum down payment of up to 20 percent. A conventional rule of thumb is that homebuyers can afford a home between 2.5 and three times the household's combined income.

- **Small- to medium-size builders** often have strong relationships with one or more mortgage companies and their loan originators, who collaborate to provide quick and customer-pleasing loan applications and processing for prospective buyers. These builder-mortgage company relationships may be either an informal or a completely formal legal partnership with financial sharing provisions in writing. Small- and medium-size builders who deliver mortgage customers into the mortgage company's hands deserve a little share of the mortgage pie as well.
- **Big builders** control the steady stream of new mortgage applicants and therefore earn the profit associated with those new loan originations. It is big business to originate thousands of mortgages a year, and all big homebuilders participate in the mortgage industry either directly by owning their own mortgage company or indirectly by way of a formal partnership with an existing mortgage company. Most of these giant builders own their own mortgage company and share the profits with no one. Big builders financially incentivize their buyers to use their mortgage company because there is profit in originating qualified mortgage loans. For example, they will pay the buyer's closing costs, **if** they use the builder's mortgage company.

Professional Designers

There are several creative design specialists involved in each step of the residential project, as follows:

- **The land developer's designers:** The design team for the land developer likely consists of a professional land planner, a civil engineer, a landscape architect, and the creative marketing team and their marketing designers.

  The land planning and civil engineering tasks may be performed entirely by the civil engineer, but often there are two distinct design teams from two firms, with one performing the concept design and the second responsible for the technical civil engineering.

  The land developer's design team(s) land plans the project, including the overall community theme and different lot sizes, and produces the engineered drawings, as required for all project approvals and construction. There also is likely to be a landscape architect involved, who designs the project's hard- and softscape elements to capture and complete the project's theme and market identity.

- **The homebuilder's designers:** The design team for the builder consists of the architect or home designer and their computer-assisted design (CAD) designers and a structural engineer, who brings the technical engineering requirements and detail to the home design.

  The initial home designer may or may not be a licensed architect. There is a distinct architectural sector found only in homebuilding; this is the home designer. These designers are often in extremely high demand, and their services are easily as expensive as a licensed architect. Their sketches are residential art that can command a premium over conventional architecture. The highest-profile home designers operate a little like famous clothing designers.

  However, if you use a new home designer, you still need the architectural seal on your drawings that only a licensed architect can provide. Many builders employ a staff architect, who manages the plan shop and CAD operators and is required to seal all the builder's drawings, even those initially designed by the contract home designers.

  The structural engineer's scope of services includes the foundation and frame design. Some municipalities require specific mechanical engineered drawings for permit application. Normally, those engineered drawings are provided by the mechanical contractors themselves as needed to pull their permits. Therefore, there are several engineers working directly for the builder's MEP contractors. Their specialty drawings complete the construction drawing sets usually required by municipalities.

  Traditionally, all the engineering services are managed by the builder's plan shop or may be part of an outside architect's scope of work.

Real Estate Agents

Americans will likely own several homes as their lives change over the years and decades. Realtors are an important link in the new homebuying and home-selling process on which the residential industry depends.

American builders have developed specialized homes for each stage of life. Professional realtors help existing homeowners sell their existing homes so that they may buy the builders' brand-new homes associated with each stage of life. Builders may not love paying commissions to realtors, but they sure love it when the Realtor walks into their model home with a live prospect!

Below are a few important notes about professional Realtors and homebuilders.

First, neither the builders nor their selling employees need to be licensed realtors in Texas. If you own the home, you may sell it yourself. Only when you act as an agent for a third-party home seller must you be a licensed professional Realtor. Employee salespeople for builders are an extension of that building company itself and do not need a license to sell that company's homes.

Second, today in Texas, builders still rely on independent agents to bring them many of their buyers for their new homes. Builders in Texas go to great lengths to encourage agents to bring buyers to their model homes.

The service of physically bringing and introducing the qualified buyer to the builder earns the Realtor 3 percent—although sometimes more—of the sales price of the new home. Note that in the post-COVID-19 housing boom, builders have pushed back on paying even the 3 percent, but as soon as normalcy resumes in the market, builders will again happily pay the Realtor's commissions.

Third, builders need real estate agents to successfully sell their clients' existing homes so that the buyers are free to buy the builders' new homes.

Some realtors offer home sellers a guaranteed buy-back on their listed home. In this way, the buyers are assured of their ability to close on their new home. That buy-back guarantees the builder of the buyer's ability to sell their current home, which is an important service to the builder and to the buyer client.

As you can see, builders need a great working relationship with local realtors to reach active home shoppers and facilitate the sale of the prospect's current home. The builder-Realtor relationship is changing due to the powerful direct marketing done by builders today. There will probably be more changes ahead for the builder-Realtor relationship, but so far realtors continue to prove their value in the new home market.

Surveyors

Did you know that the licensing process and exam for a surveyor is one of the most difficult professional exams there is? Historically, the surveyor has played a significant, sensitive, and precise engineering role in America. The surveyor was the one person with whom everyone could agree that wherever the surveyor says are the property's boundary lines, all parties will all abide by that surveyor's determination. This was and still is a huge community responsibility.

Image 1.1

Today's surveyor will perform early lot fit studies and later produce the plot plan, stake the house corners, perform the form and final survey, and provide finished floor elevation certificates and more. Use your surveyor often to avoid problems that never have to happen.

The surveyors are on your job site several times during construction. Surveyors can be your best friend when the incorrect placement of a single form board can derail the entire home project. The most common error avoided with the surveyor's help is placing the home encroaching inside the designated setbacks as specified on the plot plan. Production builders learn to collaborate with their surveyor to keep their house projects moving error-free.

Lawyers

Homebuilders will employ several types of lawyers in the course of years of building homes. Builders need a transactional lawyer to help with buying land and lots and with creating the home purchase agreements used in selling homes. If and when a lawsuit is threated or filed, a builder will employ a litigating attorney, and an environmental lawyer will likely help navigate complex environmental property issues. Also, because production builders have many employees, they also need the advice of an employment lawyer. These lawyers specialize in their categorical discipline and can collaborate as needed where events and circumstances overlap. Active production builders can expect to need advice and council on all these topics over time.

Most of large production builders have their own in-house council, who not only advise senior managers in real time but also centrally manage all the legal issues confronting the company in total. The company's legal executive manages the entire legal process on the company's most pressing issues. The general council guides and assists local operating executives and managers through difficult legal events and issues. In-house council brings together attorneys of various disciplines to collaborate and attempt solve problems as soon and as inexpensively as possible. Additionally, the in-house general council manages the company's legal budgeting and negotiating as it relates to legal expenses. The general council manages the entire legal process in large production builders, allowing operating units and their managers to continue to perform building and selling homes as undisturbed as possible.

Title Companies

What is title insurance? Title insurance provides a guarantee to the homebuyer that the seller is the rightful owner, having a good and clear title and the authority to sell the land and improvements such as a home to a willing buyer. Mortgage lenders require title insurance for new homebuyers, and builders also rely on title companies to confirm the seller's ownership before they buy raw land or finished lots.

Speaking of land ownership, when it comes to home ownership, the two most common forms of personal home ownership are *fee-simple ownership* and *community or condo ownership*. ***Fee-simple ownership*** means the owners own the actual land under and around their home, usually in the form of a platted lot. ***Condominium ownership*** is the resident's common ownership in the land and possibly also the home. This type of ownership is a little like the ownership of the right to occupy the dwelling and participate in the ownership governance. Condos are often associated with for-sale apartment-style homes of some kind, although there are all kinds of dwellings sold as part of a condo ownership. Note that when we were looking at types of homes included in residential, there was no mention of condominiums. That is because any of those homes listed could be sold as a condo ownership.

Although it may seem unlikely that a land or home seller would engage in selling land they do not own, title issues are more common than you would think. Chain of title is recorded in the county courthouses across the country. These records will extend back to United States land grants as well as land grants from the Republic of Texas and, later, from the state of Texas. Beginning with the land grants to the first individual, and every time land ownership was transferred to a new owner, a deed was recorded with the county clerk. Title insurance companies look for good and unbroken tile to the current Seller to ensure the current Buyer that the land is good to buy. And if there is ever a title claim, the title company's insurance policy will defend and/or pay off that claim against the new owner's land. That payoff is why mortgage companies require title insurance on every mortgage loan.

THE UNITED STATES OF AMERICA,

TO ALL TO WHOM THESE PRESENTS SHALL COME, GREETING:

FIGURE 1.9 The United States government made land grants directly to individuals throughout the eighteenth century in accordance with various acts of Congress to reward those engaged in specific events or service in wars. The United States land grant shown above, dated May 2, 1860, and signed by President James Buchanan, was made to a military officer who served in the Virginia Militia War of 1812. To ensure the ownership of land, one must trace and confirm the present-day seller's ownership all the way back to the land grant to be able to guarantee to the present-day buyer that the seller indeed owns the land being sold. Notice that this deed was "RECORDED Vol. 393 Page 17." The title company will be able to see that recordation at the county courthouse.

CSM students should understand what title insurance provides for the projects they will build in the future. Title companies play a key role in assuring all parties that the project has good and clear title to proceed and for all subsequent buyers along the path to homeownership.

Note that it is common practice for the seller to pay for the cost of title insurance in any land or home transaction. Additionally, the price of title insurance is set by the state, so there is no need to shop for better rates between title companies because all title policy rates are the same in Texas.

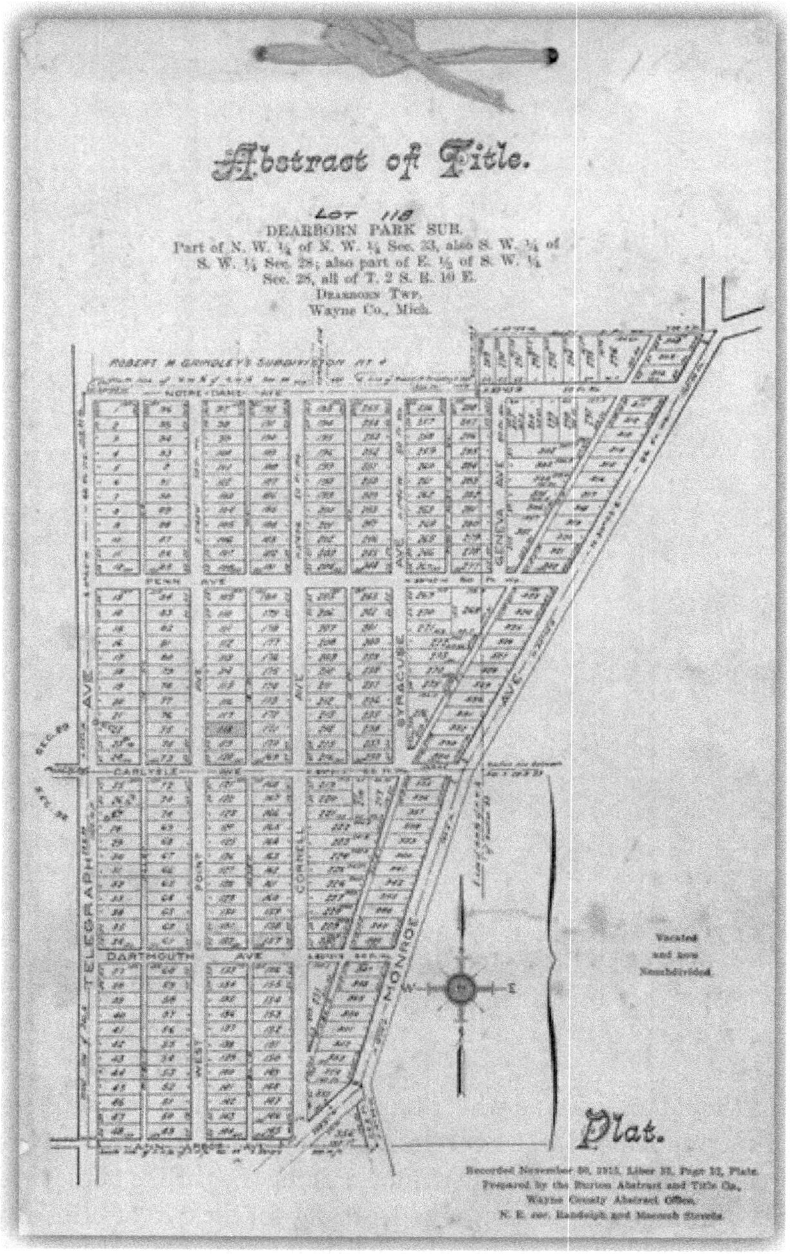

FIGURE 1.10 Title companies study the land's ownership history dating back to the land grant and compile an abstract of title document, which literally lists the sequential chain of title back to the land grant. If there are gaps or changes of ownership that are unclear due to an inheritance or a transaction with multiple owners, then therefore there exists the possibility that not all required parties were involved in one of those events. Maybe one person died, and that is why there is no signature on the deed by that owner. Most of the fuzzy periods happened long ago, and there is almost no chance of a problem. Breaks in title, if any, will become exceptions to the title policy's coverage.
However, the title company provides a list of exclusions from their title policy on items they identify as being even the remote possibility of a problem. Buyers must read and accept those exceptions from their title policy.

SUMMARY

Homebuilding is an original and essential profession in which competent constructors may enjoy great careers helping families find their affordable and perfect home for their unique family.

Homebuilding is a team activity in which you will work with many other professional disciplines. The builder's job is to ensure everyone on the team is successful. Homebuilding is a career that will call on your technical knowledge, creative talents, and interpersonal skills.

Homebuilders manage not only construction but also their entire environment and all the forces involved with their projects. Managing all of these external forces affecting their project is complex and makes homebuilding challenging and interesting.

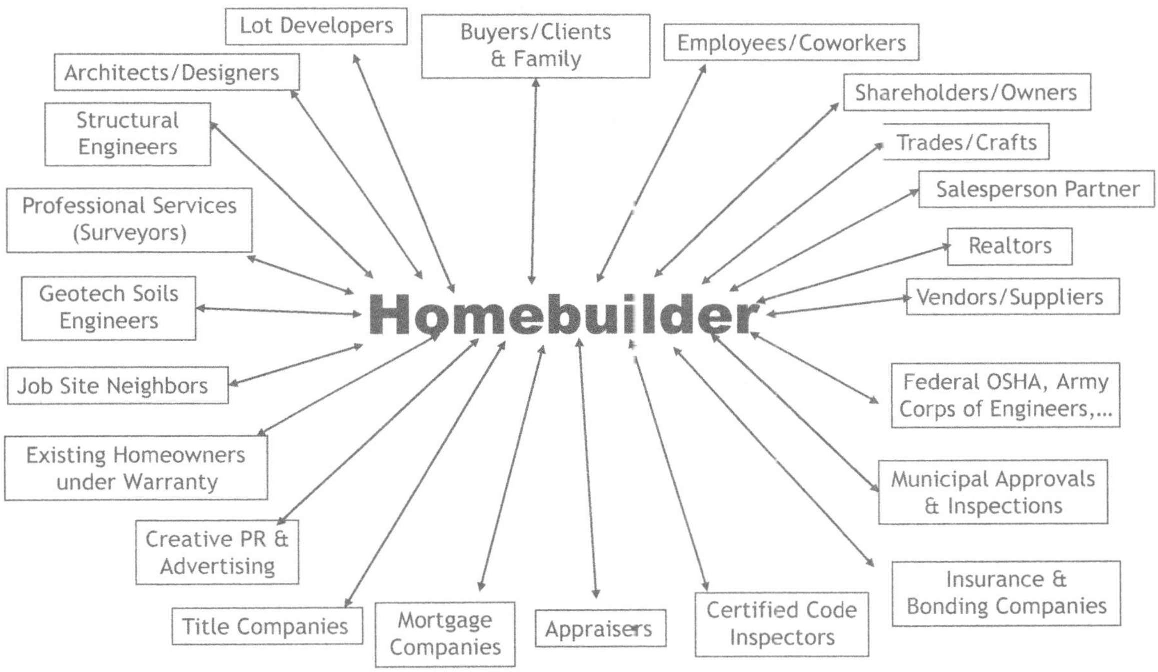

FIGURE 1.11 The life of a production homebuilder involves daily interaction with an array of people on every topic about all your homes under construction. Look at all the people and their likely topics the production builder must manage to successfully deliver their new home projects. If you love homes in general and love working with many people on complex projects, you will love building homes and making people happy homeowners for a living.

REVIEW QUESTIONS

1. What defines *residential construction*?
2. Describe the two categories of residential construction, *single-family* and *multifamily*, and in what ways CSM graduates' careers differ if they enter into either categories.
3. Describe the homebuilding profession's evolution, from the early independent craftspeople building homes, to local general contractors, to today's sophisticated national homebuilding operations.
4. How did residential real estate become such a big business? What are the three steps of the financing system that enable the industry to exist and grow?
5. Define the difference in a *public* and *private builder ownership* and the advantages and disadvantages of each. What market shares do private versus public builders hold?
6. What is the distinction between a *vendor* and a *subcontractor*?
7. How often is the surveyor on your job site, and what kinds of problem can they help the homebuilder avoid?
8. Describe the various professional designers and engineers commonly employed by residential land developers and then homebuilders.
9. What role do real estate agents play in today's new home market, and what direction or form might a relationship with them take in the future?
10. What valuable service does the title company provide, and why do we buy title insurance?
11. Why do production homebuilders often own a title company and/or a mortgage company? What is it that the builder controls that makes owning title and mortgage companies make good sense?

Figure Credits

Fig. 1.2: The National Builder, vol. 7, no. 1. 1888.
Fig. 1.4: Copyright © 1974 by Lennar Corporation.
Fig. 1.5: Copyright © 1953 by Circle Hardware Supply.
Fig. 1.8a: Copyright © by Google.
Fig. 1.8b: "Post Card," 1906.
Fig. 1.9: Source: https://www.worthpoint.com/worthopedia/james-buchanan-1860-military-land-1799554807.
Fig. 1.10: "Abstract of Title," 1915.

CHAPTER 2

CAREERS IN HOMEBUILDING

INTRODUCTION

It takes an extensive technical skill set, strong interpersonal skills, and a creative eye and mind to build homes for a living. Building homes is the coordination and administration of a series of complex construction processes that result in a beautiful and functional retail product built to last for decades and delivered personally and directly to the buyer, the end user.

Construction Science and Management (CSM) students can find rich and rewarding careers in homebuilding. Not only can they secure a great career start, but they can purposefully develop their career. Designing a career over four decades will lead the professional homebuilder through a challenging and meaningful career. Envisioning a 40-year career takes patience and discipline, but the rewards can be amazing.

Homebuilding Is Retail

Homebuilding is a retail industry. CSM students need to understand this important distinction between a residential career and a commercial construction career. Commercial construction is business-to-business. Relationships between the constructor and the client are more business and less emotional. The commercial building is built to specification and inspected objectively to meet or exceed specification by professional engineers.

Homebuilding, on the other hand, is business-to-consumer. Homebuilders build, sell, and deliver their home products directly to the end users, the happy homeowners. The specification to which homebuilders must build to is customer satisfaction, which can be challenging. The only industry more emotionally charged than homebuilding is probably child daycare management. Think about it: People are very emotional about their children; next is their house. Homebuilding requires a personal and intimate relationship with your clients, and builders must work with their clients individually to set and meet the buyer's expectations.

If you like people and you like construction, then you will love retail homebuilding.

Homebuilder Size Changes Methods of Operation

A sole proprietor homebuilder must achieve mastery of and then perform each of the complex processes required to produce even just one on-budget, on-quality home and, therefore, a happy customer. Small builders must wear more than just a hard hat; they need to wear many hats. On the other hand, medium- and large-size builders can afford progressively more sophisticated, specialized departments managing each of the homebuilding processes on a large scale. Graduates starting in small companies will be exposed to and learn different parts of the business than those graduates starting in large companies. Beginning builders can find their perfect start at small-, medium-, or large-size homebuilders.

Mentorship

CSM students will soon choose where to spend their formative, early-career years. It is not the salary or the title that CSM graduates should be seeking, but rather mentorship. The salary and titles you want will come when you are ready.

Graduates should start their careers wherever they find the best teaching system and learning environment and/or personal mentorship in construction. Join a company where you find the person who seems most interested in helping you learn and succeed early in your career. That early success will quickly build the confidence and skills that will last for your entire career. Wherever mentorship is found, this is the best place to start a career.

Earn Tenure in Several Departments

Production homebuilding is a team activity. Successful careers in production homebuilding require mastery in at least one functional department, such as construction, purchasing, or sales. However, becoming a homebuilding executive requires knowledge and competence in all the homebuilding departmental disciplines.

It is recommended to CSM students that during the first half of their career, they should seek experience in multiple homebuilding departments. Although some of those departmental moves might require accepting a temporarily lower income, nothing will accelerate a long-term career and income growth like successful tenure in multiple departments.

Great careers in homebuilding can start in any homebuilding department, including construction, accounting, marketing, sales, purchasing, warranty service, or the lot acquisitions/ development group. Spending two to five years in each of several different departments helps develop an individuals' experience base so that they can confidently manage volume homebuilding operations with their experienced base of knowledge and deliver results that will further accelerate their career.

If you want to be an executive of a homebuilding company someday, then spend time in as many departments as possible, and you will soon build a resume that will allow you to write your own career ticket.

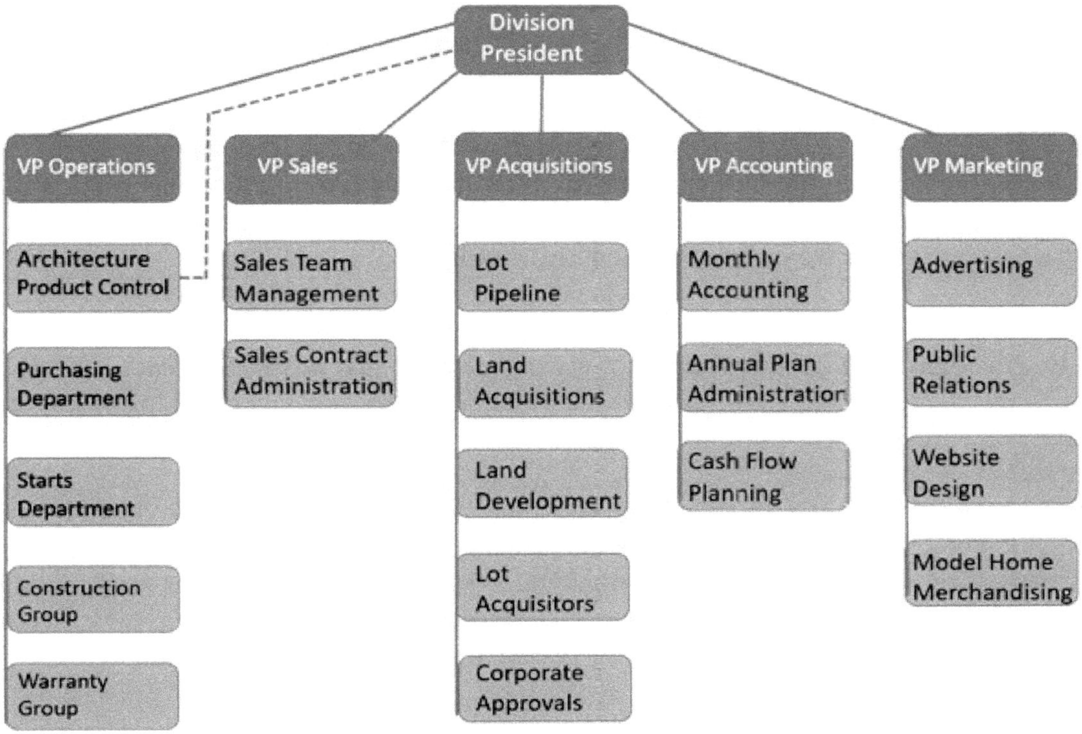

FIGURE 2.1 This production homebuilding divisional/city organizational chart shows categories of responsibility for each executive vice president position. Organizational configuration can vary significantly from company to company, but these tasks and departments must exist somewhere inside the company. Architecture and purchasing teams are often standalone departments outside the operations group.

FINDING THE INTERNSHIP OR CAREER START

Beginning a career in homebuilding can happen smoothly and without stress by following a few simple steps and rules. Start with the humble attitude of a young apprentice and be open to whatever opportunity presents itself. The steps presented below apply equally to internship or full-time career seekers.

Career and Compensation

Homebuilding is a bonus industry, and some bonus structures are designed to be a high percentage of a builder's or purchasing person's annual salary. Bonuses will easily represent 15 percent to 25 percent and up to more than 50 percent of a homebuilding professional's compensation and can range dramatically by company and by market. Builders are normally paid a per-house bonus shortly after the home sale closes for the buyers. The purchasing bonuses are typically annual events and not tied directly to any single house, as they are in the construction group.

CSM students may observe that the starting salaries in commercial construction are higher than those in homebuilding. Although that may be true, the base salary does not tell the whole compensation story. Graduates should study the entire reimbursement in total to properly weigh the opportunities. More importantly, graduates should aim for the construction sector about which they are passionate rather than for the highest base salary. If you are passionate about your industry, you will achieve your financial goals no matter which field you choose. Compensation should be low on your career-choosing metrics because if you enjoy your work, you will be promoted and earn what you deserve. Remember, mentorship is the key.

Starting at a Large-Volume Builder

Finding the right place to start means finding early mentorship to ensure your early construction experiences are successful. Mentorship and early success are the keys to launching a great career. Production builders are not only great places to start a career; they are also great places to spend a career.

For most CSM graduates, I recommend starting your career with a large production builder. Production builders have great processes, and learning those processes is a good reason to start your career in one of them. Doing so will allow you to learn their proven building methods and see successful processes operate to control and manage costs and quality. Learning how a successful homebuilding business operates is a worthy reason to start in a large-volume builder.

> High-volume builders are also the most active in hiring new graduates.

High-volume builders go to great lengths to hire and train new talent. They have robust and informative onboarding programs to bring new graduates into high-performing building teams and start their careers successfully with a lot of training and support. Professional development continues throughout your career with these high-volume builders. Bright and curious builders can mature into true master homebuilders by building homes and growing their skills inside all these successful production builders.

> Career production builders choose this path.

Most career builders simply enjoy building homes but do not want the administrative headache of being promoted to a project manager or owning a homebuilding company. If you love building houses and satisfying your customers but you do not want all that comes along with being a company manager or company owner, then building a career as a production builder is perfect for you.

> Production homebuilding is a lucrative career.

So, if you want to simply build homes or to lead the entire company someday, you will find both opportunities in production builders. Self-employment is always the production builder's biggest competitor for their best building. To keep their top employees for an entire career, production homebuilders must compete with or beat the earning power that their best builders might earn as private, small builders. Production homebuilders keep their employees with great compensation and an excellent work-life balance.

Starting at a Smaller Builder

Starting a career with a private, owner-operator homebuilder is also a great place to learn the homebuilding business. Small builders must be masters of all the departmental tasks, and therefore you can learn the entire business by starting your career in one of them. Additionally, private builders may also offer a high level of personal mentorship. Mentorship is the key to a great start.

Small builders must know a lot about everything from construction to real estate law just to stay in business, so listen carefully and learn. If you start with a small builder, take every opportunity to attend meetings with the buyers, bankers, architects, accountants, and realtors. Attend every meeting you can.

CAREER ENTRY POINTS

CSM programs prepare graduates to enter the homebuilding industry most often by way of either the construction or the purchasing department. These two departments are perfect starting points to learn the business from the field or the corporate office. Builders and warranty employees will gain important experience through working directly with the buyers. However, starting in the purchasing department will provide valuable experience from the back-office side of the business. Both are important places to start.

The Buyer Relations Experience

The single most important knowledge and skill to be acquired from one's time as a builder is the direct buyer relations experience. The career-changing importance of one's direct buyer interactions cannot be overstated. Knowing how buyers think and react to the homebuilding and warranty process will affect executives' decisions for the rest of their career. After all, homebuilding is nothing without the happy customer. Even in the construction and warranty groups, everyone is operating in a retail environment.

The construction and warranty groups continue to be a great place to launch and/or spend an entire career. Construction knowledge and experience is basic experience for a homebuilding career, so embrace whatever opportunities present themselves to you.

The Construction Group

The construction group is the most obvious place to start for CSM students. Historically, most homebuilding executives either started or at least spent years in the construction group. We are homebuilders, so logically, the builder group was the source for promotions and training the company's future leaders. The construction path is the most common but not the only first job for high-achieving construction graduates entering homebuilding.

Early production homebuilding companies promoted whichever employees could build homes the fastest and for fewer costs. Construction was messy and inefficient, so there was a lot of room for improvement. However, over time the production process improvements became less dramatic, and high-production homebuilding became the norm. Production building didn't become easier, but it did become more systematic and predictable. Now, high production and great quality are simply expected. Because fast and efficient homebuilding is the norm today, how, then, may future leaders improve their overall performance?

FIGURE 2.2 Interns and graduates will likely start their homebuilding life in a construction trailer office such as the one pictured. Builders normally work from temporary construction offices that are set up on vacant lots inside the community. The office will be located in an area of the community to be built last and then later moved to the next area of construction as needed.

Remember, homebuilding is a retail industry and team activity. Calm and collaborating communication skills are what allow good builders to become great builders.

The Warranty Service Group

CSM graduates may find their first industry opportunity working in a warranty service position. The warranty service role is an important and delicate assignment from which the graduate will learn the methods and materials of construction and contribute while learning.

Many graduates ask if warranty service is a good place to start their career. The fact is, it is a great place to start. You will get to know your trades and be exposed to the job site's environment while you perform warranty duties. Additionally, nothing teaches you more about quality than fixing poor-quality work.

Some graduates who are invited into warranty may ask, "Why was I not offered a builder job?" The answer is that in many homebuilding companies, the entry path into construction is through the warranty door. This is not a step down or an indication that the companies think you are not up to building homes. This is a training path that will help you prepare to become a builder when the time comes. And that opportunity will likely come sooner than later.

Many companies prefer their builders to start in warranty so that they can see firsthand all the problems that can result from poor-quality initial installations. If you get to see where all the problems are, perhaps you can avoid those problems on the homes you build.

Although the warranty team is normally part of the construction department, in some large companies the warranty group may be its own standalone department. Most warranty department managers started their career in construction and later found their combined communications and construction skills were best utilized in leading a warranty group. Warranty is an important group, and you may start or end your career there. It is important that students know that warranty work is important to the company and part of a good career path. Many executives started in warranty, and if that opportunity is the door that opens for you, embrace it and learn everything you can while you are there.

The Purchasing Group

The purchasing department has grown in responsibility and stature as a key department inside every successful homebuilding company. The purchasing department leader (e.g., vice president, director, manager) is normally a member of the company's senior executive team.

The purchasing role has evolved from an assistant to the builder to an estimator to a purchasing agent to a purchasing manager to, currently, a supply chain and national accounts manager. The responsibilities and influence of the purchasing department continue to increase as the group's influence reaches the company's base house specifications. Defining the company's base home product defines the company's market identity.

The quest for easier, more reliable, faster, and less expensive construction components and methods is one of the many responsibilities of the purchasing department. The home product itself is also ready for revolutionary changes. CSM science students will learn how homes are built today but should also expect big changes over the course of their career.

The purchasing department is the company's research and development (R&D) department. The homebuilding industry needs vast changes to keep up with the future demand for new housing. The future of tomorrow's home construction is unknown. It could go in many different directions, from more panelization or complete prefabrication, three-dimensional printing, or another construction method not as yet conceived of. The impending evolution of the home is exciting and part of the challenge for those students who choose homebuilding as a career.

The research and resulting evolution of the home's construction methods and components is one of the primary responsibilities of the purchasing department. This department is an integral part of the R&D team investigating new construction methods or new component products considered for inclusion in the home product. Therefore, the purchasing group is another great place to start a homebuilding career for those who want to explore and be on the leading edge of their industry.

STEPS TO BECOMING A HOMEBUILDER

Follow these six simple steps:

First, decide in which city or cities you would like to live.

Homebuilders hire directly out of the cities where their positions are open, so the first step is to choose the city or cities in which you want to live and work.

Texas is the number one state in the nation for new home starts, with about 100,000 new home starts annually. Dallas/Fort Worth and Houston are first and second in national ranking for new home starts, respectively. The Austin/San Antonio corridor is setting new home starts records year after year. There is no better place to start a career in homebuilding than in one of the metroplex cities of Texas. Be open regarding your cities of interest. Being open to multiple locations will open many more opportunities. It is likely that you can find your career start in any of cities across Texas or the nation.

Second, identify all the active builders in your chosen city or cities.

Make a list of the builders, big and small, that are building in your chosen city. Builder ranking lists for each city are compiled by various consulting companies, business media outlets, and the local builders' associations. These builder lists are available online, or you can check with the local builders' association for a membership roster. Look, and you will find a list of the most active builders in your city. Note that many of the same big builders are actively building in all the major cities in Texas, and you can apply for jobs in multiple cities at one time.

Third, individually research the homebuilding companies you found.

Zero in on the builders in which you are most interested. Read their "About" section on their website and search for news articles about the company. Look for the builders with whom you most connect and at which you would be most excited to start your career. Keep a record of what you've learned about each builder and use that information during the interviews you will soon attend.

This step of investigation and connection is important because your personal connection and positive view of that company will show itself on your first interview. Your positive mindset will get you the job offer you want. Once you identify the builders in your area that excite you professionally, you are ready for the next step.

Fourth, apply directly online for open positions advertised on the company's "Join Our Team" or "Employment" section of its website.

Today, applying for open and advertised positions online is the only way in which applications are accepted. Even if you connect with the company in another way, you will be asked to apply online for that open position.

An *open position* is a defined position that is publicly available to all candidates and for which the company is actively seeking and accepting applications from the public.

Homebuilders only hire for positions they need and have opened for applications. Mailing resumes to these companies or talking with a friend who works there is not applying for a job. Builders will not see your resume and create a position for you; they must first have the open position and be actively seeking candidates. This point is also a legal point. Hiring practices are legally trickly, so do not focus on anything but applying for actual advertised, open positions.

Fifth, research and apply for open positions on job aggregator sites such as Indeed, Monster, and others.

These sites are helpful for looking more broadly across the job market. Searching these sites with broad terms such as "homebuilding" may bring up an open position that you would not have otherwise looked for or found. This is a great way to see what else is out there besides what you found on directly on the builders' websites, so be open to various positions you might find that you did not anticipate.

Search under broad headings such as "homebuilding," not just "construction."

WATCH AT https://tinyurl.com/b3642d93

WATCH AT https://tinyurl.com/mryasf4m

Sixth, before the interview, get prepared.

Learn everything you can about the interviewing company. Visit their website, communities, and sales offices in person. Try to visit with an employee on site or in the model. Learn everything you can. Do not walk into an interview with a homebuilder and just say, "No, I have never visited one of your communities." You can see how that would come across as poor, right? If you take an interest in the company, they will take an interest in you. Before the interview, make a list of the top two or three reasons you want to be part of this company. Take that list with you to the interview and let them know specifically why you want to work for them. Do not be shy!

Application Hints

Timing Is Key!

Do not just look once and say that no one is hiring. Tomorrow, the perfect job could open in your preferred city. Persistence will pay off. All the companies on your list will have the open position for which you are seeking, but you must be looking when they are also doing so. Timing is the key. Watch constantly and respond immediately when a position opens. Students may become frustrated when they are looking at a specific company but do not see the position they want or the one they want in the city they chose. Take your time, and you will eventually find what you need.

Apply for all the open positions you find and be inclusive in your choice of positions. Never exclude a company based on something you heard or read; if it's advertising an open position, apply for it. If you land an interview, see what you can learn about them firsthand for yourself.

Preemployment Tests

During your first interview, you may be asked to take a personality or industrial psychology test. These preemployment tests are common, and the graduate should be familiar with and anticipate taking one of these tests. These common preemployment tests examine aptitude, intelligence quotient, personality type, and, supposedly, personal integrity. You may ask the company representative setting up the interview prior to the interview whether there will be an aptitude test and, if so, which type of test. You can study the question types in advance, which may help you feel more confident. An example of a common test used by homebuilders is the HighMatch (formerly Berke Assessment), and you can read more about it and take a practice test at their website (https://www.highmatch.com/).

PROFESSIONAL DEVELOPMENT DURING YOUR CAREER

Students may think that when they graduate, they are now educated, trained, and ready for life and whatever their career may bring them. The fact is that the training and personal and professional development is only just beginning. Employee development and training is common in homebuilding companies. Participating, learning, and then putting into practice the teachings from your company's training will advance your career.

Training Is a Privilege and Should Be Welcomed

Career development should be part of your personal career plan. Employee training is expensive and should be welcomed and embraced. It is also an opportunity to interact with coworkers from other departments who have various fields of expertise. Use your training time as another opportunity to network with your coworkers. Training is normally reserved for promotable, high-potential employees. Respect and appreciate it because it is a privilege to be invited.

Seek training on your company's information technology (IT) systems. Take it upon yourself to become expert in your company's IT operating systems. Learn more than just the screens required to do your job duties. Dive deep to understand exactly how the system is used by your manager and other departments inside the company. Over time, learn how your company's entire IT system is designed. Is it one integrated system or a collection of accounting, construction, and sales systems?

The best homebuilder employers—those with the greatest number of tenured employees—are dedicated to and spend big money on employee training and development. It is a compliment to be invited to participate, and you should welcome every opportunity to do so.

Company Organization and Career Paths

Volume is the single most impactful characteristic in defining how a homebuilding company will operate and be organized. Small builders evolve as their home volume increases. Individual performers evolve into a systemized collection of specialized departments. As volume grows, the small builder breaks apart work processes, assigning those parts to team members who have developed into specialists in their assigned processes, and thus the various homebuilding departments are born.

Today's homebuilding giants have grown into enormous companies that are composed of large departments performing a high volume of specialized work. Each department has developed its own performance metrics that drive future performance improvement. These high-volume, high-performing departments all need high-performing leadership. There are important careers as managers and executives in all of the homebuilding specialty departments.

The logic to a company's workflow must constantly be reconsidered and redesigned to achieve new and ever higher standards of performance. At some point in each professional's career, he or she will find the responsibility of designing the company's workflow will fall to them as they become the company's new leaders. Be studied and be prepared; your turn will come!

360-Degree Customers

Homebuilding professionals who wish to develop their career are advised to learn their company's exact workflows between one department and another. Understanding each department's role and how work flows internally will enable you to see all your personal customers: *your 360-degree customers*.

Homebuilders sometimes conduct 360-degree customer surveys around key people or positions. These interviews are designed to give the employee feedback on how well that they are working within the existing system based on feedback from all the system's participants.

Students entering any new career must understand that everyone with whom they interact is their customer. This includes the buyers, of course, but it also includes the trades, vendors, coworkers, lot developers, and professional service providers, such as surveyors and inspectors. If you strive to meet the needs of all your customers, you will find that you are working as part of an efficient and effective construction team and collaborating to deliver beautiful homes to happy homeowners.

Homebuilding is a complex process requiring a lot of interaction with many people, and your ability to communicate and collaborate with all your teammates is key to your success in homebuilding.

See yourself as surrounded by customers every day, and you will have a great career.

THE HOMEBUILDERS ASSOCIATION

The National Association of Home Builders, Texas Association of Builders, and local association chapters such as the Home Builders Association of Greater Austin and Greater Houston Builders Association collectively represent homebuilders' interest at the local, state, and federal levels.

Representing our industry's interest—or advocating and lobbying for our industry—is the essential service the associations provide. Homebuilding is a highly regulated industry, and the industry must advocate our position on important issues to preserve the availability of the American home. Beyond lobbying, the associations provide many educational and other valuable services for both large and small builders.

Professional Fellowship

There is also a professional fellowship component to participating in the builders' association. A professional fellowship among people working in the same industry and the same city. It is especially important to exchange ideas, knowledge, and solutions between professionals within the same field. The associations provide many business and social opportunities throughout the year at which attendees get to know people from other companies and departments and talk about the current market and building conditions. Naturally, we can learn from each other.

CSM graduates who wish to accelerate their career should take advantage of the training and networking available through being active in their local builders' association. Again, this advice is not just for residential students. All sectors have their own associations.

Am I an Association Member?

CSM students are likely to start their homebuilding career inside a company that is already member of the local homebuilding association. If your company is a member—and it probably is—then you are also a member of the local association. All it takes to be a member is working for a member company.

As a member, you have the privilege of attending and participating in a vast world of education, training, state and local government lobbying, and fellowship. I did not say that it is free to attend all these functions, but your company may pay for you to attend if you ask. Additionally, these functions are not expensive. If you pay your own way to an association function, consider it a small yet valuable investment in your career.

Do not be shy and think association membership is only for our company's managers and executives. You are a member just like them, and the association will welcome you with open arms. Attend every barbecue, product expo, or event you can. Go on your personal time, on weekends, and in the evenings to the events and to the meetings during lunchtime. Get involved! The association is where you will meet people from all the builders in town and grow your industry contacts and industry knowledge.

This professor recommends that everyone become active in their local association. Get active in your local homebuilder association, and watch what will happen to your career.

SUMMARY

Great careers start from a series of purposeful actions taken along the entire length of one's career. Your career starts with a vision backed up by actions. CSM students have positioned themselves well to launch a great career. You took the right first step with your education; now stay on that path by understanding and purposefully designing your career along the way. Remember that homebuilding is retail, so think like a retailer. The buyer is not your only customer; learn to serve all your 360degree customers. Homebuilding is complex. You can build your knowledge and career by working in various homebuilding departments and fully participating in all the available training.

Follow the six easy steps listed above to find the right position and company for you. Once you find your career start, get involved with your local association and meet your peers and competitors. These relationships will last for an entire career.

REVIEW QUESTIONS

1. Why is homebuilding a retail industry? What defines any company as a retail company?
2. What role does mentorship play in choosing where to best start your career?
3. It's recommended to CSM students to spend a few years in various departments, not just construction, during the first half of your career. Why is this?
4. Which are the three most likely groups for CSM students to start their homebuilding career, and why all are good career paths?
5. What is an *open position*, and why is that concept so important when actively seeking a career start in homebuilding or any construction sector?
6. How can you prepare for preemployment tests, such as the HighMatch (formerly Berke) Assessment?
7. What is the concept of the 360-degree customer?
8. What is the number one priority of all building associations, and why is it important?
9. How can you know if you are a member of the local builders' association, and how can you get involved in your association to advance your career?

THE PRODUCTION HOMEBUILDER

CHAPTER 3

INTRODUCTION

What makes a homebuilder a production homebuilder? How is it that one production builder can build eight, twelve, or more than fifteen high-quality homes comfortably and correctly at the same time? The answer is the support the production builder receives from the company's specialty departments that perform all the construction support tasks and thus allow the builders to focus exclusively on their jobsites.

Production homebuilding is a team activity. The production builder relies on teams of specialists who support the builder on many levels. It is important for Construction Science and Management (CSM) students to understand and see the people and systems around them and participate as part of a well-practiced unit.

DEFINED BY THE PROCESS, NOT THE PRODUCT

CSM students should not associate production homebuilding with a certain quality or value level of the homes being constructed. Great systems and teams allow high-volume homebuilding at almost every price point or design level. The production homebuilding process is not exclusive to any certain price point or home complexity level. Great production processes also allow production homebuilders to build beautiful, high-design, and high-finish homes.

The perception that production homes are lower-quality, mass-produced homes is incorrect. First, even the lower-priced production homes are well-designed, properly built, and completely inspected homes. Secondly, there are plenty of high-end production homebuilders. The most well-known among those high-end production builders is Toll Brothers Homes. Toll Brothers builds more than 8,000 houses a year with an average sales price of about $800,000. They build luxury homes for discriminating buyers, and they build those high-end, semi-custom homes using a production system of home construction. The product, not the process, defines the home.

Look at Toll Brothers' website and judge for yourself how beautiful, high-end homes are production-built.

WATCH AT https://tinyurl.com/yw5p9mbh

PRODUCTION BUILDERS OFFER HOMES AT EVERY PRICE LEVEL

In Texas, many production homebuilders have a long history and excellent reputation for building luxury, move-up homes. All the major Texas homebuilders build homes at all price points and value levels. It is common to find homes by the same builder in the same community but on different lot sizes, building homes priced from $300,000s to more than $700,000. Those builders build all their homes, at all price points, using a production construction system.

WHY STUDY PRODUCTION BUILDING?

CSM students pursuing a career start in homebuilding are most likely to find the first stop in their homebuilding career is with one of the high-volume production builders. High-volume builders obviously need many builders, so naturally they do a lot of hiring. High-volume builders are a great place to start a homebuilding career, and here is why:

- **High-volume builders sell a high volume of homes for good reasons.** They are successful companies that know how to operate their business, know their customers well, and build great homes that a high volume of people desire to buy. These builders have proven themselves in the marketplace. You can be proud to build these high-demand homes.
- **Production builders have well-honed operating methods and systems that you can learn.** You will apply what you learn across your entire career. As your career matures, your knowledge base evolves from the sum of all you learn at all the departments and companies where you worked. Production builders know what they are doing, and you will learn a lot by working there.
- **Production builders offer great long-term career opportunities.** Production builders are not just a place to start and learn but are also good places to build an entire career. This professor built and enjoyed a long and prosperous career working in production homebuilding. Big builders have big departments and big career opportunities.

Production Homebuilding Is a Team Activity

CSM students who have participated in and enjoyed any type of team-based activities in community service, academics, or athletics will enjoy and have great success in production homebuilding. Production homebuilding departments work to allow production builders the opportunity to focus exclusively on their homes' quality, schedule, and costs. It takes a large team playing many homebuilding positions to build production homes.

HOMEBUILDERS ARE GENERAL CONTRACTORS

Production homebuilders are general contractors, contracting out all the actual construction work to specialty subcontractors. High-volume builders normally self-perform almost none of the actual construction work using hourly or salaried crews. There are exceptions for punchout crews in some companies, but as a rule, all the physical work is performed by independent, specialty contractors. In this professor's opinion, if you need to hire punchout crews to complete the subcontractors' jobs, then what you really need are better scopes of work for your subcontractors.

Builders' Principal Duty and Responsibility

Production builders' principal duty is to orchestrate and confirm that all trades and suppliers perform their specific tasks on schedule, on budget, and according to specification. Production homebuilders are responsible for every facet of the on-site home construction project, starting from the first day standing on the vacant lot all the way to personally handing the keys to the new homeowners of their beautiful new home.

THE CULTURE OF CUSTOMER CARE

- Clean job sites
- Roofs that do not leak
- Foundations that do not fail
- New homes that do not flood

Adhere to these standards, and you'll have a solid foundation in customer care.

—Randy Birdwell, homebuilding industry director/associate professor of practice, Texas A&M University, and retired homebuilder owner and executive and past president of Texas Association of Builders and Greater Houston Builders Association

Managing the Flow of the Construction Process

Production builders must have solid knowledge of home construction methods and alternatives and all the various components and mechanical systems installed in the home. But, most importantly, production builders must know and understand the rhythm and flow of the construction process and be a competent leader for all the specialized trades during each stage of construction. Production builders must know a lot about everything in the home; however, they will never know as much about each craft as the professional subcontractor who performs the work. CSM students can continue to grow their technical skills by learning from their trades.

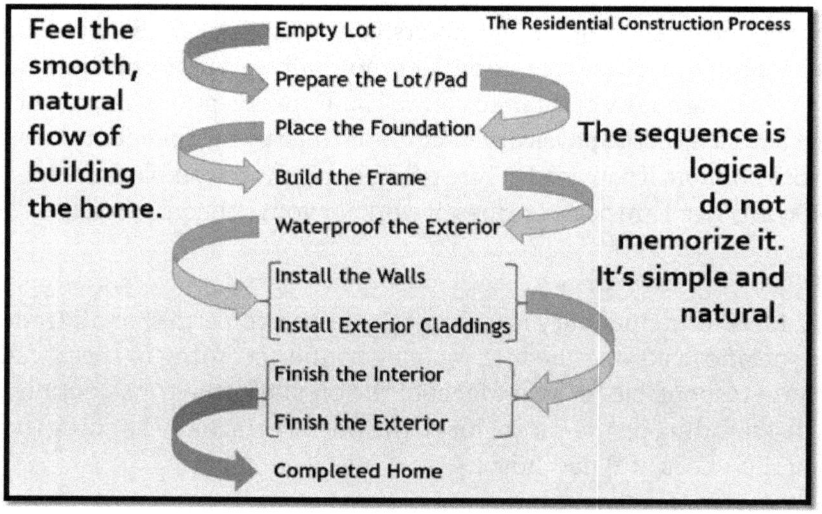

Image 3.1

Directing High-Volume Construction

The production builder has been compared to the director of an orchestra. Just as the orchestra director signals to the various instruments when to play, the production builder must signal to each independent contractor and supplier exactly when to enter and when to exit the job site. The builder directs all the subcontractors to play their part in sequence and on time. Experienced production builders master this flow and seem to effortlessly build on-budget and on-schedule homes with happy homeowners year after year.

The production homebuilder can manage eight to 16—or more—homes under construction at one time only because there exist several specialty teams using proven work systems to perform all the support tasks for the actual construction of the house. On-site builders couldn't carry a dozen houses if they had to negotiate prices on every component in the house, find next year's lots, pull permits on the new starts, work on the new marketing campaign, secure next year's financing, and meet with buyers and realtors at a moment's notice. These homebuilding specialty teams perform all the support functions that surround and allow builders to be successful, high-production, home construction contractors.

Smaller Builders Wear Many Hats

CSM students who envision themselves building in a smaller company or custom home environment will learn that smaller builders must wear many hats aside from their hard hat. Small-company homebuilding is not production homebuilding. All the homebuilding support teams and the various categories of support tasks they perform simply become part of the small builder's everyday workday. Everything you might learn from a production building environment easily translates into great knowledge and experience to the other side of the volume scale: the small, private builder.

The small builder starts early each morning by lining out the job site with the trades. Later in the morning, they meet with their accountant to be ready for lunch with their banker. After lunch, there is a hopeful meeting with a Realtor and prospective new client. Next is a marketing meeting in the office about the new website being developed, after which it is time to review and pay invoices before walking the job sites again and making the daily calls to schedule next week's work. And before 5 p.m., the builder must check the permit applications pending at the city for the job that needs to start next week. They can return phone calls in the car or later that night. This is just another day in the life of a small builder.

CSM students who aspire to someday be a custom or small builder will learn volumes of knowledge by first working inside a production homebuilder. Learning the industry from inside a production building environment will prepare an aspiring builder for a future private label building career. Where do you think all the current private builders across America learned the business?

SUMMARY

Learn everything you can every day and everywhere you go. CSM students who choose to build homes for a career will eventually learn the role of each of the support teams that are backing them up, and over time, they will learn how the existing system operates and why it was set up that way. You must become a student of everything homebuilding and learn something new at every opportunity. To become a true professional, start your career with an unending curiosity about the all the work being performed in all the homebuilding departments across the company.

As your career matures and develops, you will manage those departments and improve those existing operating systems and methods of operation, rearranging workflow to achieve all new levels of efficiency and going on to lead several complex and dedicated teams to completely new and higher levels of construction efficiency and customer satisfaction.

Understanding the various departments, systems, and workflow inside production homebuilding companies will benefit your career in whatever homebuilding path you choose to follow.

REVIEW QUESTIONS

1. What is it that allows a builder to be a production builder? Is it their product or processes?
2. What are the advantages to starting your career with a large production homebuilder?

3. What are the advantages to starting your career with a smaller homebuilder?
4. What should you really be looking for in a career start?
5. Production homebuilders are *general contractors*. What does that term mean?
6. Why might you compare a production builder to an orchestra director?
7. How do small builders, who do not have departments full of specialists supporting them, manage their homebuilding operations? What is a day in the life of a small builder like?

CHAPTER 4

THE CONSTRUCTION-SALES TEAM

INTRODUCTION

Construction Science and Management (CSM) students starting their homebuilding careers will quickly find out that they are one-half of an important partnership with the salesperson in their community. Builders must recognize and embrace this construction-sales partnership to build their earnings and career. Successful production builders are those that form a strong, collaborative relationship with their community salesperson partner.

The Community Base

Production builders sell their homes from each community's model home, their employee-staffed sales office. Each community is assigned one or more builders and one or more salespeople who all office and work exclusively in that community. Their assigned community serves as home base for those builder(s) and salesperson(s) and is the only community where they can sell or build homes.

Community assignments and teams can last for years. The success of that community is in the hands of that team. Successful builder-sales teams will work from that community base until all the lots and homes in that community are sold. Upon completing all the homes in a community, both the builder and the salesperson will be reassigned to a new community.

The Store

The community model home is sometimes referred to as a *store*. The construction-sales team runs its own homebuilding business from that store. A community selling 24 to 36 homes per year at $350,000 each is generating revenue of $8 million to $12 million annually! That is easily as big of a homebuilding operation as a small or custom homebuilder.

A well-run store delivering more than $10 million of new home revenue can be a very profitable store indeed. Each community—each store—is a profit center. A *profit center* is any portion of the builder's business where revenue and costs may accounted specifically for that operating unit and a profit and loss statement may therefore be produced. Each store has a profit goal for the year,

FIGURE 4.1 The model home park is the company's retail store and is open seven days a week; however, it is also a hub of much daily homebuilding business activity. In some cases, the construction team builds an office from the model's garage, and the model home serves as both construction and sales offices.

which the construction-sales team works to achieve. The builder's and sales' shared community goals are an important concept for CSM students to understand.

SALES BUTTER EVERYONE'S BREAD

Salespeople are compensated by either straight commission or a hybrid form of a salary plus a commission. Most salespeople are 100 percent commission, but times are changing, and new hybrid compensation programs are being tested all the time. Either way, the more homes that are sold, the more the salespeople earn in compensation.

Builders are normally compensated with a base salary plus a per-house construction bonus. In residential construction, everyone is in sales. The more homes that are sold, the more construction bonuses are paid. Ambitions, bonus-hungry builders are true partners in sales with their salesperson teammate. Builders and their partner salespeople's incomes are tied directly together in a community-based partnership focused on sales and closings, and that works to everyone's benefit.

Selling Spec Homes

About half of the homes that production builders usually construct are spec homes, or those built before they are sold; the remainder are presold homes. Keep in mind that the sooner that the spec home sells, the sooner there will be another spec home started and another construction bonus paid. For that reason, production builders are interested in robust sales of the specs in their community. Spec home starts are a vote of confidence in the community, and it is the duty of the team to sell and deliver those specs.

Successful production builders have a keen eye for how to make a positive impression on a prospective buyer visiting their community and making sure all their spec homes are always in presentation-quality condition. The weekend is the most likely time when prospects will visit a community, and the best builders always have all their homes in presentation-ready condition by that time. The builder cleans the community and job sites on Friday afternoon, sweeps the houses out, and does everything possible to make customers visiting the homes feel positive about any house they visit. Successful builders always ensure good, safe walking access to the home in case the salesperson has a prospect to whom they want to show it.

FIGURE 4.2 Spec homes are those that the builder started without a buyer under contract and are now in inventory and ready for a quick close. Spec homes are an important part of how production builders meet the needs of shoppers who prefer to see the home before they buy and buyers who need their new home quickly.

If a spec home is complete, the sales builder team works together to do what they can to cast that house in the best possible light. Perhaps additional landscaping or seasonal flowers in the beds are needed, or maybe it is necessary to fertilize the lawn, get rid of the weeds, and green it up a bit—or whatever is needed to be sure that the next prospect to visit that home is positively impressed.

Homebuilders usually do not pay the builder a bonus and never pay the sales commission until after a home is sold and closed. For that reason, both the salesperson and the builder are always doing everything they can do to make their homes look great and sell and close fast.

WORK LIFE IN THE COMMUNITY

The sales and builder team work together daily to accomplish the community goals of selling homes per the annual plan while achieving high customer satisfaction scores. The goals and compensation of the builder and salesperson are purposefully tied together.

The builder and salesperson walk houses together to resolve buyer concerns and questions or meet with buyers and/or realtors in the model home to discuss items of concern as a group. The sales-builder relationship is a little like a work marriage, their income is tied together, and they talk and work together every day and share the common goals of delivering completed homes and happy homeowners.

Monday Meetings

Community weekly team meetings, called *Monday meetings*, are usually held in every community throughout the company. Monday meetings are attended by the project manager, builder(s), and salesperson(s). The project manager will drive between several communities each Monday to attend the Monday meetings and discuss the various challenges and opportunities each community is facing that week.

During these meetings, every house and every buyer is reviewed and discussed. The house-by-house review is the purpose of these meetings.

Any communications, concerns, or questions from a buyer to the builder or salesperson is discussed in detail. Issues regarding the home's progress, issues with materials or selections, late buyer selections or missing deposit payments, quality concerns of the buyer, or anything else that might affect the home's progress or the buyer's satisfaction must be put on the table at the Monday meetings and hopefully resolved that same week. At these meetings, the strategy is to give every team member their assignments to make every week productive and one that satisfies the customers.

Customer Satisfaction

Customer satisfaction is one of the most important business metrics monitored by production builders.

Very dissatisfied	Somewhat dissatisfied	Neither satisfied nor dissatisfied	Somewhat satisfied	Very satisfied
1	2	3	4	5

FIGURE 4.3 Production homebuilders survey all buyers after move-in and collect data on a wide array of performance metrics by builder, community, project manager, and city. Photo: Wikimedia Commons.

Builders usually employ a third-party customer satisfaction consultant who offers homebuilders customer satisfaction surveying and monitoring programs. These companies not only survey your customers but also provide reports, feedback, and training designed to inform every employee of their results in several categories as well as provide coaching one on one or in groups to lift satisfaction scores across the companies. National satisfaction surveying companies provide builders with national context from which to study their results.

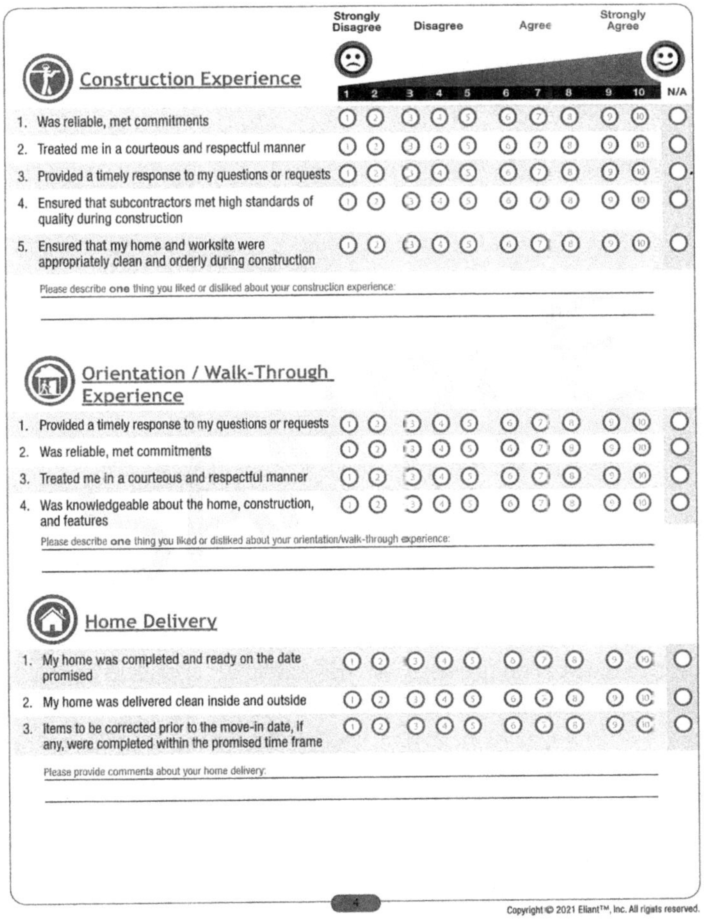

	Strongly Disagree		Disagree			Agree			Strongly Agree		N/A
Construction Experience	1	2	3	4	5	6	7	8	9	10	N/A
1. Was reliable, met commitments											
2. Treated me in a courteous and respectful manner											
3. Provided a timely response to my questions or requests											
4. Ensured that subcontractors met high standards of quality during construction											
5. Ensured that my home and worksite were appropriately clean and orderly during construction											

Please describe **one** thing you liked or disliked about your construction experience:

Orientation / Walk-Through Experience

1. Provided a timely response to my questions or requests
2. Was reliable, met commitments
3. Treated me in a courteous and respectful manner
4. Was knowledgeable about the home, construction, and features

Please describe **one** thing you liked or disliked about your orientation/walk-through experience:

Home Delivery

1. My home was completed and ready on the date promised
2. My home was delivered clean inside and outside
3. Items to be corrected prior to the move-in date, if any, were completed within the promised time frame

Please provide comments about your home delivery:

4

FIGURE 4.4 Customer satisfaction consulting companies such as Eliant and the surveys they provide (such as Eliant's survey form seen here) are the best way to obtain solid information about your buyers' experience and gain insight into your company's performance against Eliant's other builder clients. This context and perspective of your results cannot be duplicated by in-house surveys. Additionally, these consulting companies will provide valuable training for your employees and can significantly improve a company's customer satisfaction scores.

The builder-sales team is also tied together by their customer satisfaction scores. A builder's happy or unhappy customer is also the salesperson's happy or unhappy customer. Shared customers mean shared customer satisfaction results. These scores are important to both the builder and the salesperson.

In many homebuilding companies, the customer satisfaction scores are among the most highly visible and most closely watched business metrics in the company. For that reason, the harmony between the builder and the salesperson is critical to that community's success. It takes a happy builder-sales team to deliver happy homeowners.

Customer satisfaction is measured along several categories of buyer experiences. In the box below are several of the most common satisfaction metrics measured by production builders.

COMMON CUSTOMER SATISFACTION METRICS

- Was the home delivered on time?
- Did the home quality meet the buyer's expectations?
- What was the home condition at the time of the first walk?
- What was the home condition at the time of move-in?
- Were options and selections installed as agreed upon?
- What was the overall satisfaction with your builder?
- What was the overall satisfaction with your salesperson?
- **<u>Would you refer your friends and family to this builder?</u>**

The question "Would you refer your friends and family to this builder?" (or some version thereof) is probably the most universal one on all builder surveys. This metric is often referred to as the *referral rate*.

The home may have either gone up perfectly or been a difficult build; however, the bottom line is how customers feel at the end of the process. Would they recommend their builder to their family and friends? Some version of that question is the number one metric in homebuilding satisfaction scoring.

Culture of Customer Satisfaction

For some production builders, the company's customer satisfaction results are so high in the order of corporate culture priorities that no matter how your other results appear, if you do not create happy customers at the end of the process, you will lose your job. As you can imagine, tenured and successful builders are extremely picky about the salespeople with whom partner in a community. And salespeople are also just as fastidious about their builder partners.

When a community has high satisfaction scores, you will always find harmony on the builder-sales team. Where there are satisfied customers, there will also be much less stress on the team. This is not a theory; there is a direct correlation between high customer satisfaction scores and sales and high financial results in the community.

CSM graduates should seek employment where customer satisfaction is embedded in the company's culture. While interviewing, ask about the company's customer satisfaction scores. Being part of a retail industry means serving and satisfying customers; customer satisfaction is the metric on which your company's reputation is built.

Once You've Satisfied Buyers, You'll Never Go Back!

Satisfying customers is not easy, until you have done it. Once a builder-sales team has achieved high satisfaction scores, there is no going back. This is an observation this professor has made over a long homebuilding career. Once a team has achieved a certain level of customer satisfaction, it never goes back to the old habits that held them down before.

Let me give you an example. At this professor's past company, we annually selected one or more customer satisfaction categories in which our scores offered the biggest opportunity for improvement, and often these scores were stubbornly low. We revised the builders' bonus program annually to pay money directly to improve that year's chosen score(s). Not surprisingly, that bonus-targeted category's score then improved. When that score was high enough, we moved the cheese and targeted another low-scoring category.

The interesting point here that we noticed is that once our team had improved a low score, we moved the money somewhere else; however, our scores stayed high. When a builder or team learns how to do something well, it just keeps doing things properly without additional incentive.

SUMMARY

The production builder will be assigned to build in a community base and will work as a team with the salesperson(s) also assigned to that community. Community assignments for both sales and construction people can, when successful, last for many years. The builder-sales team is bonded together to accomplish the community sales and closing goals as well as customer satisfaction score goals. The team members' bond extends to a financial bond through which they can mutually earn more—or less—in bonus and commission compensation. This financial mutual bond seems to work when it comes to making a community successful!

The community hub is the model home at which many community meetings occur. The builder-sales team communicates and collaborates daily to coordinate and resolve all the issues and challenges that present themselves during each home's months of build time. Builders are often asked to join the salesperson for a buyer or Realtor meeting that will take place in the model or, in some cases, at the job site. CSM students may have envisioned their professional issues being dominated by construction challenges, but the builder's greatest challenges may come in the form of sales-related issues. The construction sales team members are financially bound also because the more homes are sold, the more homes are built, and the more bonuses and commissions are earned and paid.

Builders define *success* as a win-win with their customers, which means their customers must be happy with the outcome of their home purchase. Builders hire customer satisfaction consultants to survey, decipher, and present their findings. Production builders who cannot satisfy customers will have no place to hide. Homebuilding is retail, which means we must be customer-centered homebuilders.

REVIEW QUESTIONS

1. The construction-sales team shares common community goals. What are some of the ways the builder and salesperson have been purposefully tied together with their mutual goals?

2. Describe the concept of the homebuilding *store*. What does it mean that each community operates as a profit center?
3. Why is the success of the construction-sales team so important to the company? The team's success of also translates to the success of what else?
4. Describe the weekly and daily work routine for the builder and salesperson working as a team.
5. How does the team collaborate when it comes to satisfying their customers?
6. How is customer satisfaction measured in homebuilders? Which is the number one key metric, and why?

Figure Credits

Fig. 4.3: Source: https://commons.wikimedia.org/wiki/File:Customer-satisfaction-scale.png.

Fig. 4.4: Copyright © by Eliant, Inc. Reprinted with permission.

THE ANNUAL PLAN

CHAPTER 5

INTRODUCTION

Construction Science and Management (CSM) students must understand that every production homebuilder writes and then stewards to an annual business plan. Upon joining a homebuilding company as a builder and being assigned to an active community, you should be aware that your new community home has its own annual business plan, which now becomes *your* annual business plan. Your new job is written into that plan, and your goal is to make that community's annual plan to come true. The company's annual business plan is based on the sum of the company's individual community-by-community business plans.

Understanding your company's and your community's annual plans can help make sense of the company's current goals and initiatives. Most companies write a three-year plan, but the current-year plan is the most carefully monitored, and the present year is that which is tied to the individual performance goals of employees and managers throughout the organization. Informed and experienced builders know their community's plan goals and work proactively to accomplish the revenues from closings and gross profit margins (GPM) written into their community's plan.

Stewarding the Annual Business Plan

The annual plan plays an extremely active role in the management of production homebuilding companies. In fact, the plan is the single most important management tool used by project and area managers as well as department and divisional heads. The plan is written by these managers and approved by their managers, and it is normally presented collectively as the company's annual plan to the owners and investors. The plan acts as the preapproved annual outcome that everyone must steward month by month and community by community.

Project managers must steward their communities' actual starts, orders (sales), and deliveries (closings) against the starts, orders, and deliveries projected in the community's annual plan. Discussions at the community level are centered on the current-sales-per-month pace and the progress of the homes in the starts process and homes under construction relative to what the play says. Everyone is centered on achieving the projected sales and deliveries forecast in the plan.

Divisional or city presidents must steward the cumulative performance of all their division's communities. Collective sales are carefully watched on a weekly basis, and monthly results are reviewed in lengthy management meetings with regional or headquarter managers.

The plan is normally the basis for the incentive compensation and bonuses for the project managers, directors, department heads, and vice presidential–level employees. All managers, directors, and executives are paid bonuses to achieve that plan. Understanding this simple fact will help the CSM student understand the importance of the annual business plan.

The Bottom-Up Plan

The ideal plan is written from the bottom up, with each project manager forecasting the most likely performance of each of their communities into their community's annual plan. Each community is a profit center with its own business plan and anticipated financial results. The sum of all the company's community plans, combined with the plans written by the corporate departments, becomes the company's overall annual business plan.

During the year, the project managers must manage each of their communities individually at the community level by collaborating with the sales and builder team. The units orders and deliveries, the average sales price (ASP) and the gross profit margin for each community are clearly spelled out in its plan, by month and by quarter. The community sales and construction team must collectively steward the plan and adjust it as needed to achieve the projected financial and operating results.

CSM students should be aware of the existence of the community plan and its role in setting their project manager's goals. Additionally, they must understand the larger picture, in which their community plan is part of the company's total annual business plan. Everyone shares the same goals!

THE ANNUAL PLAN CYCLE: THE FISCAL YEAR

Production homebuilders plan and operate around their annual financial calendar cycle, which usually follows the calendar year. However, some companies follow a different annual financial and tax calendar (fiscal year), in which their year may start in November and end in October, for example. The annual plan is written forward for three fiscal year cycles no matter which fiscal calendar the company uses.

The annual step-by-step planning process is the same regardless of the company's fiscal calendar. In this discussion that follows, we use a calendar year, starting on January 1, as the basis for describing that calendar year.

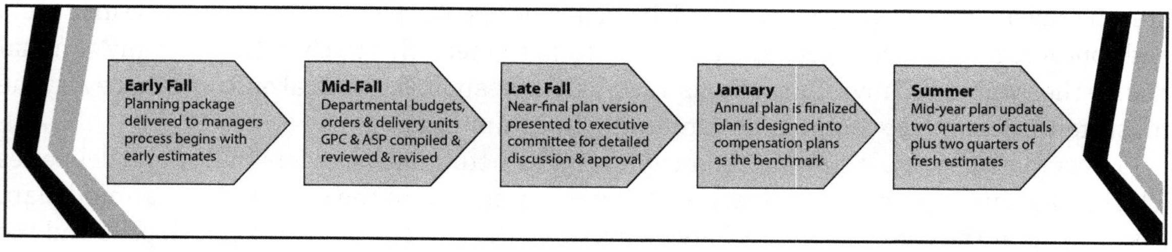

FIGURE 5.1 The planning cycle creates a highly detailed business plan annually but is also a process of constant updating and building of a rolling three-year business forecast.

Cycle Starts and Continues through the Fall

For companies using a calendar fiscal year the annual planning cycle begins in the early fall and ideally concludes at or near to the end of the year. However, if you are one of the plan writing managers, it may feel as though the cycle has no beginning or end and is simply a continuous updating process based on today's best information. A planning analyst who is part of the accounting department usually administers the planning process and provides the company's department heads and community planning teams with a planning package to be completed by specific dates.

Community Units and Expense Budgets

Upon receipt of the planning package, each community planning team writes its own plan for starts, orders (sales), and deliveries (closings), collectively referred to as the *units*. Additionally, the budgets for community-level overhead expenses are also written at the community level. Those unit numbers and community expense budgets are submitted to the planning analyst by the set due date.

Corporate Department Budgets

Corporate department heads write their departments' spending and hiring plans using the planning package provided. This package provides department managers with their historical line-item spending as a basis for next year's projections. The collective department expense budgets become the company's *sales and general administrative* (SG&A) spending plans, as written by the architectural, marketing, sales, and purchasing departments.

The accounting planning team compiles the individual community and department spending plans submitted by project and department managers into a consolidated company annual plan. The planning team assembles and compiles all the parts of the plan, analyze the results in detail, and package them all into a plan review package designed for a series of management review meetings. The plan will be discussed and revised through several iterations to become a realistic and aspirational business plan that everyone can accept.

Formal Plan Approval

Approval of the annual plan is an important and almost ceremonial management process inside all production homebuilding companies. As the plan takes final form in the late fall leading up to a series of management planning meetings are held in which the local division managers present and defend their plans' details and anticipated results. The plans' specific assumptions and operating and financial results are then scrutinized for being either too optimistic or too pessimistic. Ideally, and most importantly, the plans must be realistic.

Every company follows a different approval process, but in the end, after several iterations, a final annual business plan emerges and is approved.

Plan Updates

During the year, the planning cycle includes at least one major update based on actual results to date as well as a new forecast for the remainder of the year. This mid-year update fine-tunes the end-of-year goals for each operating unit and department. The mid-year update is not as exhaustive of a process as the fall plan writing process is because it uses hard results from the first half of the year, and only the second half of the year is reviewed and revised; it is not completely rewritten.

The plan's mid-year update is expected to be more accurate and dependable because there are only a few remaining months to forecast. Project managers are expected to be extremely accurate about their mid-year results prediction. Although the mid-year update is a time-consuming process, it provides an important road map and action plan for the year's final months.

Completing the Plan Year

Each year's end has a climactic effect on all employees of the company who feel a mixture of relief and celebration. This is because the fall months are always the biggest closing months of the year, and everyone has been working diligently during those months to finish out the yearlong team effort. The year's end arrives only after the last house of the year closes. This annual cycle builds collective experience and comradery that creates a homebuilding team and improves their future years' performance.

The homebuilding team's experience, gained by completing multiple annual plan writing cycles and delivering the planned results, builds strong team members and develops new company leaders. CSM students have much to look forward to and much to learn by understanding and experiencing the homebuilding financial year cycles.

PLANNING FOR EXISTING AND NEW COMMUNITIES

Determining Active Communities for the Year

Although most of a company's communities are already active communities entering the new year, other communities are completely new and will start during the planning year. Communities are considered active in the upcoming year only if they have spec homes or unsold vacant lots available to sell and close for that year.

Plan writing teams must write each community plan based on the best information they have at the time. Project managers must have a good sense about the market's direction of the and how it may affect houses' sale prices and the sales pace. Planning managers must also project the future lot delivery dates. Accuracy in doing so will affect the plan's ability to be achieved. The new communities will become active only if lots are delivered during the planning year. If there will be any building or selling activity in the planning year, then the community is considered active.

Projecting Units: Orders (Sales) and Deliveries (Closings)

The basis for forecasting sales and closings begins by measuring the sales and closing potential at the community level, or the number of sales opportunities that exist for the planning year. Sales opportunities as well as the inventory of sold and unclosed homes combine to define the community's total closing opportunity.

Existing Active Communities

Communities that are active at the start of the year will most likely continue to be active throughout the planning year, but some communities will run out of lots permanently and end during that plan year. Existing active communities usually have existing vacant lots on the ground, but the plan may also include some to-be-delivered lots in the next section.

Existing communities have their own current performance from which to forecast the plan for next year. Because the market is always drifting either up or down, the plan writing team must gage the market and adjust current performance slightly upward or downward to reflect the market's direction.

Past performance combined with the current market environment is the starting point for the forecast. However, the team will also consider the quality and quantity of existing spec homes and unsold vacant lots, as this current inventory is an important factor in next year's likely performance. If the community is sitting on old specs as well as some problem lots that did not sell last year, then that indicates a likely slow start for the following year. On the other hand, if the community has a new section of high-potential lots and many sought-after specs on the ground, then the plan must be written optimistically. Reading the market and inventory conditions relies on the experienced base judgment of the community plan writing team.

If a community has all its lots for the upcoming year, the task of writing an accurate plan is much easier. When new lots are expected to be delivered in the ongoing community, that is the wild card. Again, it is up to the project manager and plan writing team is to make the best possible projection about future lot delivery, and plan for the starts and sales follow from that previously projected lot delivery date.

New Communities: Opening during the Planning Year

Ideally, when opening a completely new community, the project manager and plan writing team could simply look at the lot delivery date, ASP, and GPM that were written into the new community's land/lot purchase approval package. This package is the new community's underwriting package that was approved during the land/lot purchase contract approval process. The package assumes dates based on best information from the development team at that time—and, as an aside, development teams are notoriously optimistic about early lot delivery dates.

Time passes from the community's approval date, and the project manager and planning team must make a fresh assessment of the community's lot delivery date, ASPs, and likely GPMs. The lot delivery month forecasted by the project manager triggers critical dates in the plan for starts, orders, and deliveries. Plan writing must be realistic because if the actual lot delivery date is much later than the date in the annual plan, then the goals in that community's plan cannot be achieved.

Given that lots are almost always delivered at least slightly late, it is best to assume, for planning purposes, a slightly later delivery date than that promised by the developer. For example, an April lot delivery promise should go in the community plan as a June home start, according to this professor.

Once a planning team consensus is reached on the likely lot delivery date, then the starts, orders, and delivery projections may follow from that date. Obtaining the lots on time and then promptly starting homes is the key to successfully delivering on the new community's annual plan.

PLAN COMPONENTS: THE SIX BASICS

With the planning year's active communities known, each community plan is written to include the following six basic components for each active community, commonly referred to as the "plan assumptions". :

- Starts by month
- Orders by month
- Deliveries by month
- Average sales price (ASP)
- Average gross profit margin
- Community-level expenses

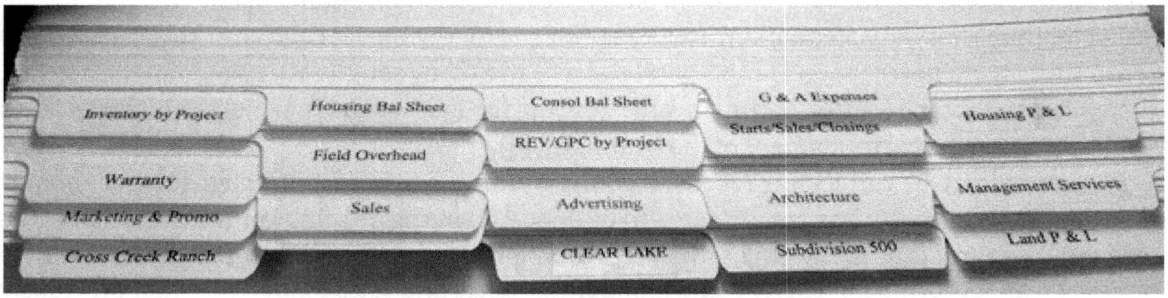

FIGURE 5.2 Once complete, the annual plan, like the printed version shown here, sets the goals throughout the year for starts, orders, deliveries, gross profit, field overhead and corporate overhead expense budgets (SG&A), and land/lot acquisition spending. The goal is for the annual plan to project and document all the year-end operating and financial results.

The initial process of community plan writing involves setting the abovementioned six variables; after this is completed, the community's plan can then be written. The community plan will provide a wealth of information, including the community's profit contribution and cash flow. The community planning process starts with a determination of the community's total sales opportunity, which is the count of every home or vacant lot owned by the builder that could possibly be sold. If the builder has only two spec houses and two vacant lots, then its maximum sales opportunity is four total sales.

Determining the Community's Total Sales Opportunities

Total sales opportunities consist of the following three components:

1. **Started spec homes available to sell and close, as mentioned above:** All spec homes are included on the sales and closings forecast counts for the upcoming planning year. Spec

homes are obviously part of next year's plan. They may or may not be complete, but they will be finished, sold, and closed in the upcoming planning year.

2. **Existing unsold, vacant lots:** These are owned or controlled (under contract) vacant lots that are available to start for the planning year and are added to the sales opportunities count. Based on the community's start capacity, these lots represent the next group of potential closings for the upcoming planning year. All the vacant lots may be added to the order count but may or may not be added to the deliveries list, depending on the community's starts capacity.
3. **To-be-delivered lots anticipated early enough to close that year:** The to-be-delivered lots are the wildcard lots. They may be added to the sales opportunities count but are only added to the closings count based on the builder's current build time in the community. If the build time is six months, then spring lot delivery might provide a few but not many extra closings for the year. Also, remember that lot developers are always late. Do not bet your annual bonus on an on-time lot delivery; always build in a few months for a future delivery. The plan writing team knows the current build times, considers the most likely lot delivery month, and writes the community's orders and deliveries from the to-be-delivered lots into the plan based on the best information provided at the time of the plan's writing.

FIGURE 5.3 Spec homes are inventory homes that were started without a buyer and are available for sale during construction process.

MAXIMUM POTENTIAL PLUS EXPERIENCED JUDGMENT

Once the total sales opportunity is established, then the planners' combined experience and judgment come into play. The sales opportunity is the maximum opportunity, but the planning team must measure that topside limit against other conditions.

After surveying the current market conditions, community's sales momentum, and current lot/home inventory quality, the team must use experience-based team decision-making to determine the best plan unit metrics for the community.

An Example

The community's total sales opportunities represent the maximum potential for new sales for the planning year. If a community has seven spec homes, 22 unsold and vacant lots (owned or controlled), and zero new lot deliveries anticipated in the planning year, then the sales opportunities (maximum sales that are even possible) is 29. Period.

Projecting new orders relies on not only the number of sales opportunities but also a reading of the current market conditions. If the previous 52-week net sales were 24, then projecting the same 24 again for the following year is easy. Projecting all 29 opportunities to sell during the upcoming planning year might be a stretch but is also not out of the question. Increasing the number of sales by 20 percent requires a plan of action. What will it take in sales concessions to increase the number of sales by five next year? Are your margins already under pressure? Can the current salesperson deliver this increase in performance? This is where the experience of past planning cycles and professional judgment come into play.

In the end, the planning team sets the planned units (i.e., starts, orders, and delivery counts), with each based on the current unit potential, current construction capacity, and experience-based market reading.

Sold Homes in the Deliveries Count

All the community's backlog of sold homes, started or not, should be included in its final delivery number forecast. If one of the sold homes cancels the contract and does not close, that house will become a spec home and will likely still close during the planning year.

Setting the Closings Count

Using your knowledge of the community's potential and market conditions, start by estimating the likely total number of closings. Then, spread the projected closings throughout the calendar year, following experience-based logic regarding when the deliveries are likely to occur. Setting target months for the deliveries in turn leads to deciding on the needed start months.

Projecting Starts Months

With the target delivery dates set and current build time known, you can determine the must-start-by months for the plan's vacant lots by simply setting the start month earlier than he delivery months by the year's agreed-upon build time, the delivery month minus the build months determines the latest possible start month for each house

FIGURE 5.4 The community's backlog of sold homes, those either complete or under construction, are the homes most likely to close during the planning year. They are always in the deliveries count.

The delivery month, minus the build months, sets the must-start-by month for each planned delivery. However, spec homes may sit unsold for a month or more before the actual closing month, so you can determine the start months by offsetting the delivery months to not only allow for build time but also provide a month or two of marketing depending on the current market conditions.

To set the community's start months, combine the likely build time and needed marketing time and then set the start month back from the delivery month by the months to build and sell each home.

Projecting Orders (Sales) by Month

The number of sales by month is not the most critical projection. Some homes will presell, others will sell during construction, and some will sell as completed spec homes. There is no experience-based way to determine when a home will sell. Therefore, planners usually spread the sales across the months in quantities per month that match historical monthly sales rates or anticipated sales rate for the upcoming year.

A word of experienced advice: It is better to backload your sales into the later months and not put too much pressure on your team in the first quarter of the year. The sales will come, and there is no need to write too much into January and February. This will give you less to explain during early monthly sales meetings.

Average Sales Price and Gross Profit Margins

The planning team must also project the likely ASP and GPM to be attained in the planning year. These two plan components are closely tied together and depend on the projected orders and delivery units.

Increasing the number of sales from those of the previous year often suggests lowering gross profit margin expectations. Additionally, improving the previous year's gross profit margins implies increasing the ASP. Doing so usually causes a slowdown in sales. Can you increase sales without negatively impacting margins or vice versa? Everything is connected!

The planning process at this point assumes the delivery unit goal is fixed, so now the exercise involves adjusting the ASP and gross margins as needed to achieve the unit delivery plan. Based on the previous year's actual ASP and gross margins and the new home market conditions, the planning team sets the ASP and margin goals for the community.

Sales prices are always creeping up just to keep pace with increasing construction costs. Cost pressure is never-ending, as even the lots escalate in price, and the entire market will respond with at least slightly higher sales prices. It is therefore normal to assume some level of price increase from the previous year.

GPM is probably the most uncontrollable and unpredictable variable in the mix. The best predictor of future GPMs is the immediate past. When the market is moving up or down, the GPM is the first variable to reflect that change in the market, for better or worse. The planning team must read the market and determine where to set next year's target gross profit margins and average sales price for each community.

THE COMMUNITY GROSS PROFIT MARGIN AND PROFIT FORMULAS

The Gross Profit Formula

The community gross profit margin formula is basic Homebuilding 101 and is the formula that sets the community's profit goal. The formula is as simple as it is important.

THE COMMUNITY GPM FORMULA

Deliveries × ASP × Gross Profit Margin % = Community Gross profit Contribution (GPC)

For example:

If you plan on closing 29 homes at 18 percent GPM with an ASP of $435,000, then you can determine the community GPC with only these three numbers.

(29 closings x $435,000) × .18 = $ Gross Profit Contribution
$12,615,000 × .18 = $ Gross Profit Contribution
$2,270,700 = Gross Profit Contribution

If the company has 10 communities, each making $2,270,700 GPC, then:

Total Company Gross Profit Contribution = $22,707,000

The Profit Formula

Once the total company's gross profit contribution is established, the sum of the company's department budgets represents its overhead budget, also known as *SG&A* (Sales and General Administration). SG&A represents all the fixed costs required for the company to operate each year. Most SG&A spending originates from the corporate office departments, such as purchasing, warranty administration, sales and marketing management, land and lot acquisition department spending, legal spending, and management and executive salaries. The profit formula is also simple.

THE COMPANY PROFIT FORMULA

Pre-Tax Profit = Gross Profit Contribution – Overhead (SG&A)

Company overhead is often referred to as *SG&A*, which was defined earlier. SG&A spending pays for the corporate office salaries and expenses and legal, marketing, and other general overhead costs required to run a company.

For example:

The company has 10 communities, with each making $2,270,700 GPC.

	Total company GPC	= $22,707,000
–	Company overhead (SG&A)	= $11,500,000
	Pre-tax profit	= $11,207,000

MANAGING THE COMPANY WITH THE ANNUAL PLAN

The role of the plan in managing the production homebuilding company cannot be overstated. It is the number one tool used by company executives to navigate the company during the year.

Homebuilding uses the principle of management by exception. The plan's financial and operating goals are the benchmarks against which actual spending and any exceptions are measured and managed. Everyone is stewarding the same plan. These common goals are the key to successfully leading a synchronized team toward a common objective.

The sales and closing unit metrics and financial results established in the annual plan are also the results against which annual bonuses are measured and awarded. For this reason, every project manager and department head is keenly watching monthly results and adjusting the strategy as needed to make the annual plan come true. The annual plan is the common goal and guides company leaders throughout the year.

SUMMARY

The annual plan is more than a financial plan; it is the company's annual operating and action plan. The plan not only sets year-end targets for orders, deliveries, gross profit margin , and profit goals; it also sets benchmarks for the operations side of the business.

The plan is the most universally used management tool in homebuilding, serving throughout the year as a guide for preapproved actions, such as buying lots and land, and as a month-by-month performance benchmark. The plan unites the entire company around specific common goals, tying everyone together with a shared stake in the year's outcome.

REVIEW QUESTIONS

1. How is the plan used to unite the entire company behind the same set of goals and precisely outline the month-by-month actions required to achieve the plan's goals?
2. What is meant by a *bottom-up plan*?
3. Describe the workflow throughout the year in the annual planning cycle.
4. What is meant by the term *active community*?
5. What is the big difference between in the process of projecting starts, orders, and deliveries in a new community and projecting those of an existing community?
6. Describe and demonstrate an understanding of the six components of the plan.
7. What is meant by *sales opportunities*, and how is their number different from that of projected deliveries (closings)? How are the two numbers arrived at, and what do they mean?
8. Describe and demonstrate an understanding of the gross profit margin and profit formulas.

CHAPTER 6

THE LAND AND LOT ACQUISITION PROCESS

INTRODUCTION

Homebuilders live a life in pursuit of the next great community to carry their company into the future. The better their communities, the quicker they sell, and the sooner they must be replaced. Pursuing new land and lots in homebuilding is a pursuit for business survival because homebuilders are out of business as soon as they are out of lots.

Production builders are staffed and organized around building a large volume of homes. Their business model is designed to perform at its peak when the system is full of houses at every stage. Gaps in new lot deliveries cause serious problems and inefficiencies and can even result in the loss of key personnel.

Land and lot acquisition is a topic so broad, deep, and important that an entire semester could easily be devoted to the subject. This presentation on the topic is designed only as an introduction and overview of the process and the key ingredients that influence the acquisition decision-making process.

Lot Pipeline and New Communities

Production builders seek to own or control by purchase contract a minimum of 24 months of lot inventory based on their current rate of sales. This means that if a production builder is currently building 500 homes per year in a city, then that city division needs to own or control at least 1,000 lots—and preferably more. These owned and controlled lots are referred to as the builder's *lot pipeline*, and it is shrinking every month and must be continuously replenished

The homebuilder building 500 homes a year is probably active in about 20 communities. It is reasonable that during each year approximately five communities (a 20 percent increase) complete construction on all available lots, thus moving off the active community roster. Additionally, this means that another five new communities need to be onboarded just to replace the sales from the closed communities. Production builders are constantly lining up new projects to replace active communities that run out of lots. A builder's new community count is an important business metric for production builders and a highly watched metric by Wall Street for public builders.

Each public builder's lot pipeline in years is another closely monitored business metric. In some supply-constrained markets, builders maintain as much as a five-year lot supply. The longer the lot development time in each city, the longer the builder's lot pipeline needs to be in that city. In most cities in Texas, a new community's inception to first lot delivery time is approximately two to three years. Lot delivery times can vary dramatically from municipality to municipality, but relative to the rest of the country, Texas enjoys very speedy lot delivery times. Therefore, it is reasonable for a production builder in Texas to maintain only about 24 months of lot pipeline.

Community Pursuit

Each city division of large production builders has a land and lot acquisitions manager who is responsible for continuously pursuing, evaluating, and filtering every possible land and/or lot opportunity identified in that city market. The land manager is always a senior member of the division management team and has extensive experience in land and lot procurement. The division's land person is critical to the success and survival of the builder's division. A continuous flow of high-quality lots is the life's blood for every builder, and the acquisitions manager is the tip of the spear in that pursuit.

The division's land person collaborates with other department heads, such as those of purchasing, construction, and sales, to ensure that current and accurate costs and sales data are built into the management teams' land and lot purchase decisions. This individual leads the land and lot pursuit, but the entire division management team must agree on each prospective new community as the acquisition decision process unfolds. The team's collective experience makes the vital decisions about the product, pricing, features, and all the other project assumptions, which then are dialed into the purchase equation.

Consensus and Commitment

The decision to sign a contract and enter a new community is a team decision, and all senior managers need to be enthusiastically on board with the decision. Onboarding new communities is a huge effort that crosses all department lines and relies on the support of all team members. As the select few communities make their way through the acquisition process, either everyone is on board, or the community will lose momentum and be dropped from consideration.

THE COMMUNITY SELECTION AND UNDERWRITING PROCESS

Filtering the Opportunities

In large cities, there are often 15,000 to 30,000 single-family starts per year. Sifting through the available projects and vacant land presented to the builder in all stages of land purchase or lot development means there is a gigantic universe of opportunities that must be evaluated to select the few opportunities worth spending the company's limited acquisition time pursuing. Builders do not have unlimited time to evaluate every deal they see; they must rely on experience-based judgment as to which new community opportunities best match the company's homes and company value proposition and are most likely to provide the financial results required.

Birddogging lots and land is the responsibility of every member of the senior management team. Department heads are normally active in the local builders' association, and networking is just part of the job for company department heads. Everyone listens for even a rumor of a new community that might be of interest, each of which is passed along to the acquisitions manager to filter through to find the few opportunities that best match the company's goals.

The acquisition manager and division president are in almost daily discussions about the communities under active consideration. It is common for there to be a monthly meeting during which the acquisition manager presents to the other senior managers the status of all community candidates under active consideration. Those meetings with the management team and the ongoing discussions with the division president guide the acquisitions manager regarding which opportunities are the highest priorities and should get the company's attention.

Value Proposition

Value proposition speaks to the unique characteristics of any product or service that distinguishes it from other similar or direct competitors. Each new community has its own unique value proposition. The value proposition may be that the lots are very cheap, and you will be able to sell affordable houses in this community. That is a legitimate value proposition and is one end of the spectrum. Another value proposition might be a community in which lots are big, wooded, and near a lake, and although they are expensive, they have characteristics that are in high demand. High price and high value are also parts of a legitimate value proposition.

A new community's value proposition is its market strategy designed around maximizing whatever natural characteristics are available to that community. Each community is evaluated on its value proposition relative to any other negative value characteristics it may have. The community may be behind a shopping center on the freeway, but it also has walkable proximity to everything people want, such as shopping, work centers, and great freeway access. Each community must build on its best characteristics. It's however you want to look at it. One person's freeway noise is another's quick access to work. Each community must choose its value proposition strategy, and the homes must match that theme.

In the end, the community's proposed value proposition must be considered as it relates to homebuilder's desired building position relative to their home products. Does the proposed community's value proposition match the builder's home value proposition? If so, that is normally the first hurdle in the filtering and selecting of the new communities the acquisitions manager will pursue.

LETTERS OF INTEREST AND INTENT AND THE PURCHASE CONTRACT

Letters of Interest and Intent

When the acquisitions manager identifies prospective raw land or a community being developed that may be of interest, that manager will reach out directly to the owners or developers and initiate the conversation to learn more and express interest in the property. Hopefully, the discussion will progress positively, and the manager will learn enough about the seller's needs and desires to translate those findings into offer to buy, sometimes called a *letter of interest* or *letter of intent*

(LOI). The LOI is delivered to the land owner and is a common tool to formally initiate a discussion about the purchase of the property.

The two types of letters are similar but slightly different in detail and intensity. Both types express the desire to buy the property, and both are legally nonbinding for either party. The difference is in the letters' detail and strength. A letter of interest is delivered in the early stages of the inquiry process and normally does not include specific details about the proposed purchase; it simply states that the buyer wants to talk. A letter of intent is more detailed and outlines specific prices and closing dates and whatever other details are relevant to the purchase. A detailed letter of intent may be signed by both parties and written with such precision and detail that it can serve as the contract outline used by the attorneys drafting the land or lot purchase contract documents.

The Purchase Contract

The contract document is usually drafted first by the seller and then reviewed by the buyer. These contracts are important documents and must be understood completely before signing. Usually all contracts are reviewed by a transactional attorney for legal implications contained in the contract. However, only the builder acquisitions team can read the business deal points contained within the contract document. The lot or land purchase agreement goes through several revisions and iterations before finally becoming a mutually agreeable and signable contract document.

Production builders often build their homes in large master planned communities (MPC) where numerous other production builders are also buying lots from the MPC developer. When building in the large MPCs, builders are allowed to make few or no changes to the contract document because all the participating builders buy lots with the same master contract agreement. The MPC lot purchase agreements are normally very one-sided in favor of the developer. This is because MPC lot positions are highly coveted builder positions, and all builders are in line to sign that one-sided contract and build in the MPC.

The contract document examination and approval process may involve various sets of eyes on the contract document, depending on the production builder's acquisition protocol. Production builders always have several projects under review, and each must follow the builder's purchase document review protocol. Finally, after a signed contract is in hand, the next phase of the new community approval process begins: the underwriting phase.

ACQUISITION UNDERWRITING

Due Diligence

Prospective new communities that advance to the point that they have a signed purchase contract now enter an in-depth investigative process known as the *due diligence period*. During this period, usually 60 to 90 days, the land or lot buyer may investigate the property and, at the buyer's option, cancel the contract within the study period without penalty or loss of the deposit based on any findings the buyer deems important. The due diligence period is also sometimes referred to as the *free-look period*.

During this period, the lot buyer investigates all aspects and characteristics of the property, including the title history, soil conditions, environmental history and condition, and home market and economic condition, all of which play into the purchase of the land or lot. Additionally, on raw land, the builder buyer contracts with a civil engineer to apply civil construction costs to a proposed land plan, adding complexity to the project investigation process.

During this period, not only are the physical characteristics of the property evaluated, but also the economic and financial conditions of the project's surrounding market are studied. Often a general market study is performed by a third-party consultant to confirm the viability of the proposed homes and retail pricing strategy. The market data uncovered by the market study feeds into a financial model, the project pro forma, which must also be developed during the due diligence period.

Community Plan Assumptions

Plan assumptions are the basic building blocks that build a pro forma community business plan for each community under consideration. The financial results from that pro forma point to the final acquisition decision in a buy or don't-buy direction. The assumptions that the acquisition team must derive are literally the factors in the equation used to build the community's pro forma financial projection. The basic assumptions used in the pro forma must be agreed upon by the acquisitions team building the community's financial model.

BASIC ASSUMPTIONS THAT BUILD THE COMMUNITY'S PRO FORMA

- When will the lots most realistically be delivered?
- What home products (i.e., floor plans) will be offered?
- At what specification level (high or low) will the plan offerings be made?
- What construction costs can be anticipated at the time of lot delivery?
- What will be the average sales price (ASP) based on the home and specification level chosen and the delivery date?
  - » What gross profit margin will be the result of those assumed cost and price assumptions?
  - » What build times can be anticipated at that time and on those specific types of lots?
  - » What sales or absorption rate can be anticipated given the assumed sales prices and lot delivery date?
  - » What special costs, such as model home or spec home requirements, are required in this project?

Internal Rate of Return

Production builders normally use the internal rate of return (IRR) as the financial hurdle for each new community's decision-making. The IRR is the cash-on-cash financial return relative to the money invested over the life of the project.

IRR is a common financial metric that indicates the discount rate at which the project's net present value equals zero. Although that explanation leaves a lot of explaining for a future construction finance class, IRR indicates the profit returned based on how long the cash is used by the project. The financial pro forma uses these basic assumptions to build this financial model, which will conclude with an IRR buy or don't-buy indicator.

The longer cash is used in a project, the more profit it needs to generate to pay for the time during which that cash was at work. For this reason, the later cash goes in and the sooner it emerges, the better the IRR. For example, if the builder must close on the land and then spend a year on approvals before construction may commence (as opposed to if the land purchase could be postponed for that year during the approval process), that factor would make a huge difference in the IRR earned by the project. Time is money, and IRR measures both the amount of money and the length of time for which that money is invested.

Each project must clear the company's financial IRR hurdle rate to proceed in the approval process. A large production builder sets various hurdle rates in different years to accommodate different housing markets. In difficult markets, builders are normally willing to accept a lower hurdle rate, and in a good market, they may hold the standard to a higher return threshold. The key is to set the current rate to match the available market opportunity.

The bottom line is if the company's IRR hurdle rate is 18 percent and the project's pro forma turns in a 14 percent IRR, then changes to the team's basic assumptions—or a change to the land price—needs to be made. If the team is fixed on their basic assumption and the seller is unwilling to renegotiate the land or lot price, then the community will be dropped from active consideration. Clearing the company's presently set IRR hurdle rate is the financial metric required by most production builders' community approval processes.

The Final Acquisition Package

Each step in the due diligence phase builds the contents of what will become the final acquisition package, which serves as both a decision document and a detailed business plan for the community. The acquisition package serves as the approval document for upper management to review and is used in a series of land approval meetings.

The land and lot acquisitions package must be complete in every way according to the builder's exact form and protocol. For large production builders, this usually involves presenting the package in advance of the land meeting for senior management review. After which in the land meetings the package's contents are produced into a formal presentation form that thoroughly covers every point required by the process.

The Formal Approval

Every step in the land or lot acquisition process is highly detailed and sequential, and it's a process that is made more difficult because it must be performed inside the due diligence period allowed in the contract. The contract provides a specific number of days to investigate the property and get the deal approved by whatever process the buyer is required to follow. It is common to request an extension (or two) to the due diligence period to secure final approval of the contract and purchase deal. When extensions are requested, it is common for the seller to request an additional deposit as evidence of the builder's continued commitment to the project and process.

Formal approval is usually given first verbally in a land or lot acquisition meeting and later completed by signature of all approving managers at the division and corporate levels. The approved acquisition package memorializes the basis and basic assumptions by which the decision to enter the community was made. This package is used again upon lot delivery as an action plan to achieve the closing and financial goals set out in the acquisition pro forma. It is used once more after the entire project is complete as part of a look-back to reconcile the projected performance against the actual project performance.

The formal approval of the new community serves as the authority for the division president to proceed to close on the property and as a guide to the entire division team to move forward with the steps to open the community. The acquisition package spells out the entire project plan, from product to timing to pricing. Although this plan changes between approval and lot delivery, the community's sales and profit goals do not.

SUMMARY

Disciplined land and lot acquisition protocol is basic practice in all successful production building companies. Lot replacement is a nonstop process! The acquisition process is led by a competent, proactive, and experienced manager who heads the land and lot pursuit and study along every step in the acquisition process.

Builders use letters of interest to initiate land or lot purchase discussions and letters of intent to finalize those discussions with land or lot sellers. Identifying prospective new communities is the responsibility of all senior managers; however, negotiating terms and prices with land and lot sellers is a delicate process that is led by the company's acquisition manager.

Large production builders are constantly reviewing land and lots, and each builder has their own criteria and financial return hurdles established to guide the city managers through and to the successful acquisition of new replacement communities. The division president and acquisition manager must lead the never-ending pursuit of new land and lots to keep the division in business.

DISCUSSION QUESTIONS

1. Describe the concepts of the land/lot pipeline and the new community count. Why are these business metrics so closely watched by production builders?
2. Describe and define the concept of *value proposition*. How does a community's value proposition factor into a builder's decision to pursue a new community?
3. Describe the difference between a letter of interest and a letter of intent. How are they different, and how are they used differently in the acquisition process?
4. What is the due diligence period, and what purpose does it serve in the acquisition process?
5. What are the basic assumptions the team must set to write the project pro forma?
6. What is the IRR, and what role does it pay in the project evaluation process? Describe your understanding of the IRR and the concept of the hurdle rate.

CHAPTER 7

RESIDENTIAL LAND AND LOTS

INTRODUCTION

It takes a house and a lot to make a home, as every new home needs a new lot. Homebuilders must buy either already developed lots or raw land and then develop their own lots. The residential lot is as much a part of the industry as the home itself. Construction Science and Management (CSM) students who like the residential industry but also lean toward civil construction can find a solid career in residential lot development by working inside a big production builder or merchant lot development company.

Homebuilders need tens of thousands of newly developed lots every year in every big city in Texas. In this chapter, we will introduce the basics of buying finished lots and developing residential land into platted lots in a community.

Buying Developed Lots

Builders in Texas may purchase fully developed lots directly from land developers. The availability of developed lots—and even the existence of a separate residential lot development industry—is unique to Texas and only a few other states. This standalone lot development industry is a big advantage for small or young homebuilding companies that do not have the financing or in-house development team to buy and develop raw land.

Although buying developed lots may relieve homebuilders of the effort and risk of land development, homebuilders must still be competent in the land development process and in managing the land development supply chain for their future homesites. Homebuilders are continuously buying new lots and must be sophisticated lot experts. They need a predictable pipeline of new lots to supply their business going forward.

The lot purchase transaction is a complex business matter that is covered in detail in subsequent business practices and business law courses. Additionally, there is a technical side of purchasing lots, in which civil construction science skills are used. The builder that buys the lots must be able to properly define exactly the physical and municipally permittable condition on which both parties can agree and the language that defines a completed lot, or one ready to be purchased. Over time, homebuilders that never develop a single lot still must become professionally competent in the science of lot development.

Lot Development Is Civil Construction

Lot development is a distinct and separate construction project from that of the home; however, the home and lot are inseparable. Constructing the lot is as much a part of the residential industry as is building the home. Construction Science and Management (CSM) students considering a career in civil construction will find that a high percentage of their civil projects will be residential land development projects or otherwise indirectly related to those projects. The community itself must be planned, and the lots must be constructed from the raw land before there are any homes or office buildings, shopping centers, or schools constructed.

Residential lot development is such an important and complex civil construction activity that it deserves its own semester-long course. Nationally, one million to two million lots are developed per year. Texas is the national leader in new home starts and lot development and usually develops more than 100,000 new homesites per year.

FIGURE 7.1 Lot development and homebuilding go hand in hand. Many homebuilders are land developers. CSM students entering civil construction as a career will likely work on many residential projects.

The residential land development industry must convert hundreds of thousands of acres of farm or native pastureland into fully serviced homesites every year. Additionally, the lots are only part of the land development projects. These lots exist inside beautiful, new residential communities that are full of amenities, including lakes and drainage and utility infrastructure, which are all built as part of the overall communities' civil engineering project.

Because Texas needs about 100,000 new lots per year, and each acre yields about two-and-a-half lots, that translates to about 40,000 acres of residential land development annually in Texas. Students interested in a land development career should keep residential careers and projects in mind.

Lot Development Risk

Risk is an important concept for CSM students to understand. Residential lot development, when compared to building a spec home, carries a much greater amount of risk exposure.

First, the process of lot development involves buying raw land and developing homesites, which takes at least two years and often much longer to complete. Time is risk. The longer you hold the land or lots, the longer you are exposed to market changes. Lot developers and homebuilders are always trying to predict the ups and downs of the new home market. Building a home takes about six months, which is one-quarter or less than the risk of the time risk exposure involved during lot development.

Second, if the market turns down on the lot developer, there is little or no utility or possible income for a vacant lot while the developer waits for the market to improve. On the other hand, if the homebuilder gets caught in a downturn, a vacant home has a lot of utility and possible income. Builders can almost always sell their unsold homes at cost and get out. And, if worse comes to worse, homebuilders can rent out a completed home or even move it themselves. There are many ways and a lot of utility in the homes allowing homebuilders to sell or rent and improve their cash flow during a downturn.

There is less risk in a home versus raw land—period! This is an important concept for CSM students to understand during their developer, constructor, or homebuilder career.

SUPPLY-CONSTRAINED AND SUPPLY-RICH MARKETS

Supply-Constrained New Home Markets

In supply-constrained markets, such as those along the West and East Coast, there are few merchant lot developers—those that develop and sell lots but do not build homes. Almost all the lots are self-developed in house by big homebuilders that need the lots for their own companies' construction. Lots are of such rarity and value in these constrained markets that the only dependable way to obtain vacant lots is to develop them in house. And if you are a builder that owns lots, you would never sell them, because they are extremely difficult to replace.

Therefore, the vast majority of lots developed in a supply-constrained market are developed by the homebuilders themselves. Additionally, the few merchant developers that exist in this type of market are giant development companies, such as the Irvine Company in Orange Country, California. Buying developed lots in supply-constrained communities such as the Irvine Ranch is a difficult or even impossible proposition. These communities can name their own terms and price and can handpick their builder clients. Obtaining vacant, permit-ready lots in a supply-constrained market is exceedingly difficult and expensive if not impossible.

Homebuilders need a reliable source of new lots every year. The only way for them to ensure that supply in this type of market is to buy the raw land and develop their own lots, which is precisely what they must do. In fact, homebuilders operating in this type of market have a lot development team in house, working down the hall from their homebuilding operation. CSM students interested in civil construction as a career should keep in mind that there are great careers in land development inside many of the large production homebuilding companies.

Supply-Rich New Home Markets

In supply-rich markets such as Texas, lot development and homebuilding exist together but as two distinct halves of the same residential company. CSM students should view these two sides of the residential coin as twice the number of opportunities. Both are important residential career paths: building homes or residential lots and communities. The Texas market for lots is the country's largest and therefore the leading job market for civil construction science graduates interested in lot development.

FIGURE 7.2 The Katy Texas Map above uniquely illustrates the comprehensive way residential land developers will blanket new communities over new regions of large metropolitan cities like Houston. Each of the twenty-six named communities on the Katy Map represent new residential subdivisions being developed at that time by residential land developers. Note that all available land is consumed by these new communities. Additionally notable on this is map is the fact that the 26 communities shown in this well-known and popular "Katy" area of Houston are, in fact, not located in the city limits of the city of Katy at all. The greater suburban west side of Houston has attached the popular city's name to the entire region which helps sell new homes in all twenty-six of the communities in this map.

MERCHANT AND IN-HOUSE LOT DEVELOPMENT

Merchant Land Development

Lot development and homebuilding walk hand in hand, but in Texas, they are two distinct industries. Lot developers exist separate from any homebuilding company in the form of merchant lot development companies.

There are many lot development companies building residential lots exclusively for sale to builders in their new projects in every city in Texas. Because of the independent lot developers,

Texas homebuilders can operate profitably and grow their companies without self-developing a single lot. This unique characteristic of the Texan market offers a variety of construction careers for CSM graduates.

Merchant developers' business model involves speculatively buying raw land, working with cities to gain development approval, and physically installing the required improvements, including roads, sewer, water, and electricity, after which the developers sell the completed lots directly to locally active homebuilders.

Advantage to Start-Up, Small, and Custom Builders

The existence of a standalone land development industry in Texas represents a large advantage for any start-up homebuilding company or Texas builder simply looking to grow its company without owning any raw land. The constant flow of new and growing homebuilders ensures an eager finished lot market for these merchant lot developers.

Custom homebuilders normally do not develop any of their own custom-level lots. The lot development industry often targets custom builders as customers not only because they always need finished lots, but also because they can pay a lot of money per square foot for their lots. Custom homebuilding lives or dies by buying high-end finished lots, and only rarely do they develop their own lots. For this reason, the custom homebuilder group is an important customer for merchant lot developers.

Developing lots is a profitable business, but in Texas it is not mandatory to build new homes in new communities. That begs the question of whether to develop lots or not to? Developing lots—or not—is a strategic business decision, not a requisite of homebuilding. Below are some of the common considerations to determine whether to buy or build your lots.

WHAT MOTIVATES A HOMEBUILDER TO DEVELOP RAW LAND OR TO BUY FULLY DEVELOPED LOTS?

- Lot supply is plentiful; there is no need to take the development risk.
- Lot supply is so low that builders must develop to obtain even one finished lot.
- Lot supply is tight, and builders want to control their own lot supply, so they develop.
- Some builders always avoid the risk of raw land and lot development, period.
- Some builders want to cut the developer out and make their profit, too.
- Some builders do not have sufficient finances for land/lot development.
- Some lot developers become builders only to build their way out of their projects.

In-House Lot Development

Many Texas builders both develop their own lots and buy finished lots from merchant developers. This opportunistic approach to securing finished lots is common in Texas. Texas homebuilders that develop lots in house have a land development team that manages all the company's in-house lot development. CSM students interested in a lot development career should investigate the opportunities with large production builders that have an in-house land development team. You will find open positions in land development; they need you!

It is also common practice for builders in Texas to develop lots and sell half of their lots to a competitor builder. In doing so, the builder avoids the market risk of holding too many finished lots at one time and for too long. Builders that sell lots to other builders in Texas usually sell their lots only to competitor builders that also develop lots and will someday soon sell them half of their developed lots. This informal practice of developing and selling half the lots to competitors probably sounds odd to West or East Coast builders, but it works very well in Texas. Sharing the risk of owning lots while seeding the future for other finished lots that would otherwise be unavailable is how homebuilders collaborate in buying and selling lots between competitors in Texas.

The primary reason for in-house lot development by homebuilders in a supply-constrained market is to furnish their own internal homebuilding operations. They generally do not sell their lots to other builders. If they take the risk to develop the lots, they want the lots all for themselves. They are homebuilders first and lot developers second.

CSM students interested in residential lot development will find a great career in either merchant lot development companies or with large homebuilding companies.

Community Building—Almost Anything You Can Imagine

Building streets and utility services for the residences, schools, churches, and commercial sites connected to a project is a complex civil construction project of the highest magnitude. Community land development also includes the construction of various amenity components, such as recreation and welcome centers, parks and playgrounds, sports and tennis courts, and swimming pools and splash pads. Fully servicing the community likely involves off-site improvements and possibly even building brand new sewer and water facilities. The range of civil construction and styles of residential projects you are likely to build during a community development career is endless. Let your imagination run wild about a new community theme, and you will probably build that concept someday.

Every Project Is Unique

Every site has its own set of unique conditions and challenges that may involve all kinds of other civil construction components. Community land development often includes off-site work to connect the community to existing utility facilities and/or roads. Large projects may include construction and installation of the community's own sewer and water utility plant. Delivery of these essential services to the lots is a masterful engineering and construction undertaking.

Residential land development is a great career choice for CSM students interested in civil construction. You can build a career working on the civil subcontractor side of the civil construction or inside the development company, where you will supervise these subcontractors.

THE BASIC MERCHANT LAND DEVELOPMENT PROCESS

1. Identify and contract for the purchase of the raw land.
2. Identify the likely homeowner demand type anticipated for that site's location.
3. Obtain preliminary engineering and design services for the project at large.
4. Land plan the track to match the identified and targeted homeowner demand type.
5. Evaluate the project based on all the discovered facts before closing on the land.
6. Explore financing options and requirements as needed.
7. Shop the project to potential partner builders that might buy lots if desired or needed.
8. Make a go or no-go decision on the project and then close on the land.
9. Start selling lots to other builders or keep them for in-house building.
10. Finalize all financing as needed.
11. Complete the engineering and design of the project.
12. Obtain all approvals and entitlements from governing authorities.
13. Manage the land development bidding process to qualified subcontractors.
14. Award contracts for the construction to match the project budgets.
15. Manage the entire construction process through to project completion.
16. Create and manage a homeowners' association (HOA).
17. Finish all municipal requirements to release of all performance bonds.
18. Deliver and close on the permittable, finished lots to the builder client or internally.

BASIC COMPONENTS OF LAND DEVELOPMENT

Land development is a category of construction that cannot be given justice in a single chapter. The following is only a simple overview of a topic that deserves an entire semester.

The basic land development components included in all residential land development may be loosely grouped into three categories: dry utilities, wet utilities, and streets and paving.

Dry Utilities

Dry utilities include electrical lines, natural gas, phone, and other communications-related services such as cable, fiber optics, and digital subscriber lines for internet. Each of these utility services is installed separately by a different contractor, but each utility is usually installed in the same trench as the one used by all. Installation depths and spacing specifications vary by municipality and local utility providers and are indicated on the plans.

Image 7.1

Wet Utilities

Wet utilities include storm drainage systems, sanitary sewer collection, and potable water distribution. Each of these systems requires its own specific engineering drawings. The goal is to design and build an all-gravity-controlled storm and sanitary sewer system, which can be a significant engineering challenge. Storm water control is always run by a gravity system, which requires precise engineering design and installation to avoid future flooding issues.

Image 7.2

The storm collection and detention—or retention—is designed to do more than simply remove rain from the community. New communities are developed from open fields into subdivisions, where as much as 50 percent of the land is covered by impermeable surfaces such as concrete. Therefore, the drainage system must capture and detain the rainfall runoff, limiting the rate of the runoff after development to be equal to or lower than the rate of water runoff prior to development. The runoff limit requires the construction of storm water detention or retention basins inside the community. These can be wet detention basins such as lakes or dry detention basins that look like soccer fields when it's not raining and lakes when precipitation is falling.

Again, the sewer system design goal is an all-gravity-controlled system. Sewer collection systems can usually be designed as entirely gravity-controlled; however, in communities in which the topography and site conditions do not allow for this type of system, the solution is to install one or more lift stations to reelevate the waste to an elevation at which gravity can take over.

Street Construction

Street construction is much more than pouring concrete such as for a large parking lot or driveway. The streets are not merely driving surfaces; they are also a big component of the community drainage system, acting as large gutters. The volume of water the streets hold, from top of curb to top of curb, is included in the overall community drainage volume calculations. Streets must conform to a precise design for width, depth, elevation, and slope.

Image 7.3

There are all kinds of street designs and types of paving surfaces used in residential land development. In some cities, asphalt is common, whereas in other cities, concrete streets are the norm. There are pros and cons to both, depending on the climate and road design.

ENGINEERING THE LOTS

The community development engineering plan also includes a design for each new section of lots. That lot design may be for the lots to be left wooded and completely natural, or they may be designed as a benched or an engineered lot with a precision 2 percent grade forming the perfect A-type lot, with draining from back to front. Engineered lots are cheap to build on but usually expensive to buy. That is because engineered homesites save builders money by setting the forms without any extra lot-related grading or other lot preparation costs and using their base slab budget. Buying pricy engineered lots may result in a net savings to builders over native lots, which require expensive earth work.

On the other hand, on wooded or native lots, builders can incur all kinds of extra costs when preparing a lot prior to building the home; however, they may get paid a lot more for that home. Both lot design approaches have advantages and disadvantages, and both have important implications to the homebuilders that are buying and building homes on those lots.

One section of lots may be left completely native to save the existing trees. Another section may need to be raised by 18 inches for drainage and so that group of lots will be filled, compacted, and graded to a precisely engineered elevation. When buying lots, builders must look at the total net cost of building on each lot and factor in all associated costs of the home. Production builders eventually become competent at building on all types of lots.

Different Lot Types Need Different Solutions

Every lot condition is unique, and special consideration must be given to the best approach for each individual lot or group of lots regarding the final home's cost and esthetics. Builders want wooded natural lots for discriminating buyers who see the value of the additional costs associated with preserving and working around the trees. They also want economical lots with more predictable costs for their lower-price-point homes. In the latter case, an engineered lot may be better.

Lot Conditions and Extra Costs

Engineered Lots	Standard house budgets will work – No grading or slab extras!
Flat/Low Lots	Build an elevated pad with compacted fill creating lot drainage around the house and to the street.
Falling Lots	High rear slab – Dropped Brick Ledges, Steps, Patio Railing, Erosion control at rear
Rising Lots	High front slab – Dropped Garage, Dropped Brick Ledges, Steps, Porch Railing
Corner Lots	Extra landscaping and sod, irrigation zone, sidewalk, extra fence and more...
Amenity Lots	Extra landscaping & tree requirements, iron fence - lake lots, greenbelt lots, "view lots"
Oversized Lots	Extra sod and irrigation zones and fence -- "pied-out" lots or just big lots
Wooded Lots	Tree removal/protection, elevated pad construction, fill & compaction tests, haul off spoils

Image 7.4

In real life, community lot design includes a mix of various lot types as the project topography and conditions change in various areas and over the projects' life. CSM students who build homes will quickly learn the various costs associated with each type of lot.

Common Lot Types Faced by Homebuilders

Engineered lots: The developer grades the lots at a perfect 1 percent to 2 percent grade, one usually falling from the back to the front of the lot, toward the street.

These lots are free of native trees, boulders, and other costly conditions. Engineered lots do not require extra slab costs or extra grading budgets.

Flat/Low lots: The building area elevation is below or near the curb elevation.

The low lot requires importing compactable fill to build a well-elevated building pad with compactable fill material in 8-inch lifts and raise surrounding portions of the lot as needed to drain around the house and complete the drainage pattern depicted on the plot plan.

Falling lots: The elevation of the lot is lower in the back than at the curb.

The falling lot results in a high rear slab condition and one or more of the following:

1. Dropped brick or stone ledges, 2. steps and railing at the back or side patio, 3. extra stormwater pollution prevention plan erosion control at the rear of lot, and 4. high slab extras as prescribed in your slab contract.

Rising lots: The elevation at the curb is below that of the remainder of the lot.

The rising lot results in a high front slab condition and one or more of the following:

1. Dropped brick or stone ledges, 2. steps and railing at the front porch, 3. dropped garage and associated cost of framing and garage steps, and 4. high slab extras AND dropped garage extras as prescribed in your slab contract.

Corner lots: The lot's location is situated where the home's side elevation faces a street or community pedestrian path entrance.

The builder guidelines require additional features on the side of the home to make the visible side finished and consistent with the rest of the community, including extra landscaping and sod, irrigation zone, sidewalk, fence changes, and more.

Amenity lots: These are designated lots that either face, are behind, or are adjacent to special community features such as a lake, trail system, or park.

High-visibility lots adjacent to the community common areas require special attention by the developer to achieve a higher level of home finish including extra landscaping when visible from public community views.

	The builder guidelines require the builder to install additional landscaping or a backyard tree or special fencing and other budget-impacting costs.
Oversized lots:	These are either cul-de-sac lots with big pied-out backyards, extra-wide lots, or any lot configurations resulting in an above-average amount of land.
	Oversized lots are popular, and lot premiums can be charged to buyers to pay for the extra costs associated with the lot size.
	Extra costs include additional sod and irrigation zones and fencing.
Wooded lots:	These lots involve the preservation or removal of native trees.
	Wooded lots are highly sought after, and people love to buy and pay lot premiums for them. However, they require special precautions, costs, and efforts to prepare the building pad to successfully build a new home and can be extremely expensive.

Special Considerations When Building on Wooded Lots

Protecting trees intended to be saved is an ongoing effort during the entire construction period. Saving trees requires planning and expenses and can be a sensitive matter to both the buyers and the trees themselves. It takes a lot of effort to build a home around a delicate old tree without killing it. However, you still must connect to the sewer and water lines, which involves trenching directly across the tree's root and drip line. You must try to save the trees, but you cannot guarantee success.

Removing trees is also costly. The building pad area needs to be clear of trees and their stumps and roots. Excavating and removing the stumps and logs from the site is followed by replacing compactable fill in 8-inch lifts and recompacting and testing each lift in the affected areas to achieve 95 percent Proctor density. There are no exceptions to this protocol. The builder must be confident using test results that the pad is properly compacted before placing the home on the lot where trees previously existed.

Pad Building on Wooded Lots

Cutting or filling a native wooded lot to build a house pad in the middle of the trees you just saved involves numerous challenges and extra costs. You must ensure that no fill from cutting the streets was placed on the lot to build the lot up higher than the curb. Wooded lots often require the same extra cost associated with a flat or low lot and building an elevated pad. The pad construction complexity is compounded by performing this work while also protecting the trees.

BUYING LAND AND LOTS

Homebuilders are constantly in the process of buying land or finished lots. The ability to identify and contract to buy lots is imperative for any homebuilder. Knowing the language of lot and land buying and selling is especially important. CSM students should familiarize themselves with the common jargon used between lot developers and sellers and homebuilder lot buyers.

The following is a glossary of common nomenclature for buying and selling lots and is provided as an overview to the terminology and terms common to the land/lot purchase process.

Nomenclature Used in Buying and Selling Land and Lots

Amenity lake: A lake that has no detention capacity or freeboard and is designed into the community only for aesthetics (i.e., to look good) and for the ability to charge lot premiums.

ARC approvals: The Architectural Review Committee's (ARC) authority to review and approve or disapprove plans and color and materials selections planned for new homes; this is often granted in the deed restrictions. The ARC's reviews may be part of the land developer's team or an extension of the HOA.

Base lot price: The price of a standard (i.e., nonpremium) lot in the pool of lots being purchased by the production builder. Developer lot premiums are added to the base lot price.

Builder guidelines: The developer's contractually mandated, predesigned, allowable home construction materials, colors, and styles and the landscaping and amenity lot specifications that create the community building standards.

Builder partners: The two or three builders buying lots, building, and selling in the same section or lot program. For example, all the builders on the 55-foot lots are builder partners, as are the builders on the 60-foot lots.

Building setbacks: The required distances from the front, sides, and rear property lines from which any new structures or other specified improvements may be constructed. Setbacks can be seen on the plot plan and are usually a standard distance for most lots in a given community.

Bulk lot purchase: All the lots included in the contract are taken down all at once upon completion. There is no taking down on a schedule or rolling options on the lots over time.

Closing; to close: The event during which all transactional documents are signed and the money is funded to the seller; also known as the *settlement* *To close* is a verb that describes buying and paying for the property; *the closing* is the actual event.

Community common area: A community's HOA-owned land inside master planned communities. Common areas include the landscaped sides of collector roads, the pool and clubhouse, and roadway medians. These are the areas that the HOA must maintain by mowing, irrigating, and doing whatever additional means are required.

Contract default: The breach or failure to perform on the terms of a contract; a legal term. If a buyer does not close on the lots as agreed, that buyer is in default. If a seller fails to deliver the lots by the prescribed date, the seller is in default.

Contract remedies: A contract clause describing an agreed-upon predetermined outcome in the event of default; a narrative of what will—or can—happen should one of the parties in the contract go into default. The remedies clause and terms of default are two of the most important clauses contained in any land or lot purchase contract.

Deed restrictions: Legally enforceable rules placed on building types and land uses that are allowed on the land or lot, which always apply to all subsequent owners. These are legal instruments recorded at the courthouse along with the deed and always following the title.

Drill site: The designated location for subsurface mineral rights owners to access their minerals from below the surface; often disguised as a community park but can be legally drilled upon at any time. Drill sites exist in communities all over Texas. Legally, the surface rights owners must provide the mineral rights owners access to their minerals, even if that access is right in the middle of a residential community.

Due diligence period: The days allowed per the contract to investigate and physically examine the property before the earnest money is at risk; also known as the *free-look period*. For example, a builder may deposit $100,000 and has 30 days to perform due diligence on the property. If the builder discovers a problem with the property on the twenty-ninth day, the builder can cancel the contract in writing and receive the $100,000 back. The purpose of the due diligence period is to allow the buyer reasonable time to investigate the land or lots.

Earnest money contract: A lot or land purchase agreement that allows the buyer to exit the contract and lose only the amount of the earnest money deposit. The risk of entering this contract is limited to the amount of the earnest money. There is no specific performance for the buyer. This type of contract is the opposite of a specific performance contract because the buyer cannot legally be forced to complete the purchase.

Easements: Deeds that record the right of access and maintenance granted to various public and utility entities to allow for future maintenance or repair of specific utilities provided in the community. These are always indicated on the plot plan and are not subject to altercation or variance.

Entitled lots: Land that has acquired the required governmental approvals to develop X number of lots; however, those lots are not yet constructed, and the land is still native. These are also sometimes called *papered lots*. Raw land might sell for X amount of money; entitled land sells for X amount more. Usually, land is entitled for a specific number of lots.

Freeboard of a lake: The depth from the top of the lake water's surface to the top of the bank of the lake. During a storm event, the amount of freeboard is the distance to which the water can rise without going over the lake's banks. The amount of freeboard designed for a lake has implications not only for drainage but also for aesthetics. The less freeboard there is, the better-looking the lake is; also, the less freeboard there is, the less rain the lake can hold. The amount of freeboard is a design and engineering decision.

Land ownership; fee-simple and condominium: *Fee-simple ownership* implies single-entity ownership of an identified piece of land or platted lot. In fee-simple ownership, a buyer owns the lot and the house. *Condominium ownership* implies that the land on which the house sits is jointly owned by all the community homeowners.

Letter of intent (LOI): A written, nonbinding letter to a land or lot seller stating that intent to buy the land or lots. This letter is considered stronger than a letter of interest; however, it is still nonbinding for both parties. These letters are usually extremely specific and detailed. The LOI from a buyer is often used by the lot developer as evidence of demand in the developer's financing package.

Letter of interest (LOI): A written, nonbinding expression of interest from a want-to-be land or lot buyer to purchase land or lots from the owner. The LOI may be general and noncommittal or specific and formal. More formal LOIs may be used in a precontract stage or when purchasing a complex property for which the LOI serves as the outline of a future contract. The LOI from a buyer is often used by the lot developer as evidence of demand in the developer's financing package.

Lot drainage type; A-, B-, or C-type lots: *A-type lots* drain to the front from the back; *B-type lots* drain to the back from the front of the lot; and *C-type lots* are crowned and drain in all directions.

Lot fit study: A prepurchase examination of each of the lots contained in the proposed purchase contract to evaluate its size and shape relative to those of the planned homes. The study is designed to ensure the usability of every lot in a section for the anticipated floor plans.

Lot premiums and fees: The additional charges assessed in the lot sales agreement for various premium values or extra costs associated with each individual lot. Lot premiums are assigned to individual lots by the developer based on a share of the likely premiums the builder will charge homebuyers for that lot. Developers want a share of the revenue earned on oversized, corner, or view lots of any kind. If the builder can get more, the developer also wants more. Fees are also charged by the developer for architectural reviews or any other administrative cost that can reasonably be passed along to the builder lot buyer.

Lot program: A group of lots that are the same width and targeted to be sold at the same general price point to the production builders buying those lots in the section. Communities typically develop lots in five-foot-width increments, such as 55-foot, 60-foot, and 65-foot lots. Builders in each lot program are direct competitors and are each sold half the lots to build the similarly sized and similarly priced homes.

Merchant developer: A developer of lots for the exclusive purpose of selling them to third-party homebuilders. Merchant developers do not build any homes; they only develop lots. If they do build homes, then they are builder developers.

Metes and bounds: A precise and complete land survey of land that can become a legal description for the purposes of a sale of a yet unplatted piece of land.

Municipal utility district (MUD): MUDs are developer-created tax districts established to build and support community-specific infrastructure such as water and sewer plants and for the construction of some types of collector roads in areas in which no municipal utility services are available.

No better terms: This contract clause specifies that the developer will not offer any other builders a better price or better terms on comparable lots than those prices and terms under which the developer is selling the lots in this contract.

Phase I and II environmental: Environmental studies performed during the due diligence period to examine the land's historic and current environmental condition. A phase I is always done, and a phase II is performed only if called for by the phase I. Lenders always require a phase I report.

Phase of lots: A group of lot sections that are designed to be installed in nearly the same frame. Usually each new phase requires the extension or significant infrastructure such as collector roads and utility mains. Phases contain sections, sections contain blocks, and blocks contain lots.

Plan spacing: Contractually prescribed number of lots that must be left between similar floor plans and similar elevations of the same floor plan. The builder guidelines may specify, for example, that no floor plan may be bult on a lot within five lots of another lot where that same floor plan is built, meaning that the builder must space five lots between similar floor plans.

Plan elevation: The front view of the floor plan. Usually, each plan has multiple front elevations to allow the same popular floor plan to be built multiple times in a section without the plan's frequency being obvious from the street scene. Different front elevations change the look of the home by using different materials or completely different home styles and allow builders to push the plan spacing requirements on their popular homes. The side and rear elevations usually remain the same on all lots.

Plat: The recorded document subdividing land into numerically identified lots, blocks, and sections. A platted section of lots is a group of lots that are designed to be developed at the same time. The section is further broken into numbered blocks, and the blocks are broken into numbered lots. The subdivision name, section, block, and lot number may be used as a legal description for the sale of a lot in the community.

Plot plan: The drawing of a single platted lot that shows any existing improvements and all planned home, garage, flatwork, easements, fence, and other anticipated improvements.

Pool lots or hold lots: A lot developer might contract to sell most of the lots to the builder partners on a given lot program but hold back a few lots to be sold later to the fastest-selling builder in that program. Alternatively, when both builders sell at the same pace, the hold lots are split and sold, with half to each builder, at the end of the section contract. This contract tool is used to even out lot supply between two builders in projects in which one may sell and run low faster. The fastest-selling builder gets the pool lots.

Public improvement district: An agreement between a landowner and the city to allow development of a specific tract of land outside the already provided development patterns and provisions normally allowed by the municipality. Examples of changes might be to increase density in one area and provide public parks in another area of the planned community.

Rolling lot option contract: Finished lots are sold to a builder under a contract by which the builder must close and pay for the lots according to a predetermined schedule. The rolling option contract implies that this is an earnest money contract with only a money deposit at risk until all the lots are closed. This is an option contract, so there is no specific performance.

Sales "bust-out" or cancellation: A term describing when the buyer of a home backs out of the contract for any reason. *Bust-out, or cancellation, rate* refers to the percentage of all sales contacts written and signed that ultimately do not close. Developers always require builders to report sales and cancellations every week to keep track of lot demand and remaining supply.

Sales net, net sales: The number of total homes sold minus all cancellations during a specific period of time, such as a week, month, or year.

Section of lots: A specific group of contiguous lots of similar size and shape. A section of lots is usually designed with enough lots to match that lot program's anticipated home sales for one year. If each builder is likely to sell 24 homes per year, then a two-builder section would ideally contain 48 lots.

Seller's representations: A narrative description covering specific categories of contractual representations about the land or lots being sold. These usually cover items such as the lot" environmental or financial condition (no liens, etc.) and are written for the seller's protection. They are disclaimers.

"Sold": A home that is under contract to a buyer. A builder describing a community with 10 houses under construction might say, "We have six solds and four specs," for example.

Spec home: A home built on a speculative basis in anticipation that it will later sell. This type of home is started and built initially without a known buyer.

Spec requirement: The contractually required spec home count by the developer of each builder. The specified number includes the unsold homes that are either under construction or completed within the community. Developers want home inventory in their community and will sell lots only to builders that can afford to constantly maintain spec homes in the community.

Specific performance contract: Allows the seller to legally enforce, by a lawsuit, the purchaser to proceed and close on the land or lots. The seller essentially sues the buyer for the sales price of the land. This is the opposite of an earnest money contract.

Square footage caps and floors: Lot developers may contractually specify the size range of the homes that are allowed to be built on the lot. The contract will state that any home bult on this lot will contain a minimum of 2,800 square feet and a maximum of 3,400 square feet of living area. This home size constraint is enforced in the developer's plan review.

Substantial completion (date): The narrative description defining a completed lot as a clause in the lot purchase contract. This definition spells out the exact conditions that define lot completion, thus triggering other time-critical contract dates. Normally the builder receives a letter from the developer's engineer declaring the exact date that the lots were substantially complete. The builder has the opportunity to contest in writing that the lots are not substantially complete for reason X, and the developer has a specific amount of time in which they are allowed to respond. Otherwise, the date of the engineer's letter will become the substantial completion date. The substantial completion letter gives formal notice to the buyer that the lots are ready for purchase according to the terms and schedule in the contract.

Takedown schedule: The contract exhibit identifying the purchase timetable for the contracted lots to be purchased and paid; the dates for each "taken down" per the schedule.

Terms sheet: An offer description from a merchant developer (i.e., seller) to homebuilders (i.e., buyers) explaining the lot product being offered and the desired lot prices, closing schedule, and other business deal requirements of the prospective land or lot sale.

Total lot price: Includes the base lot price plus any lot premiums and lot developer–assessed "fees" charged. Fees and premiums are not small amounts, so be sure you understand how the total lot price will be calculated at the time of the takedown.

True-up: Developer sells the lots to the homebuilder initially at the base lot price, but when the home closes to the homebuyer the builder may need to make an additional payment causing the total lot price to be equal to a specified percentage of the home sales price. The true-up payment amount is calculated to make the final and total lot price equal to a specified percentage of the final home sales price. For example, if the developer wants 20 percent of the home sales price, and the base lot price turns out to be only 19 percent of the final home price, the builder must true up and pay another 1 percent after the house closes and the final sales price is known.

SUMMARY

In conclusion, residential land development blends civil and residential construction together, thus offering CSM students a broad and varied career in any of the multiple disciplines involved with residential construction.

Residential land development is part of the homebuilding industry and can successfully exist inside and outside of homebuilding companies. It is also a retail business that requires the developer to design and develop new communities precisely for the end user, the home buyer. Although this chapter does not do justice to the broad and deep subject of developing residential lots, it does introduce the subject and invites students to learn more.

REVIEW QUESTIONS

1. What are the advantages and disadvantages of buying fully developed lots rather than developing land yourself?
2. How are land development risks and homebuilding risks different?
3. What are the various reasons that motivate homebuilders either to develop lots or to not develop their own lots?
4. Describe the distinctions and characteristics of buying lots in a supply-rich market and buying lots in a supply-constrained market.
5. What is the difference between a merchant lot developer and a builder-developer that develops lots?
6. Describe each of the three major components of land development.
7. Describe the various lot types and the common home construction costs and implications to homebuilders based on lot type.
8. Define important land and lot nomenclature and the implications of each to the builder when buying land or lots.

Figure Credit

CHAPTER 8

LAND PLANNING

Vision and Design Process

Kerry R. Gilbert and K. Michelle Stoops

INTRODUCTION

Land planners play a unique role in the real estate development process. Their primary role is to produce ideas and concepts for a project that are typically illustrated in the form of sketches and drawings. These are often referred to as *plans* even though in every instance, they are handed off to an engineer or architect to produce actual construction plans that incorporate the regulatory standards necessary to obtain approval of the specific project and enable it to be constructed. Simply put, land planners serve the role of an artist who produces concepts and ideas for land development projects. They also assist their clients in determining proposed land uses and development densities (number of lots, etc.) and advise as to regulatory processes and time frames for obtaining approvals. Land planners may also meet with various governmental authorities to obtain their input and coordinate the approval process.

OPPORTUNITIES AND CONSTRAINTS

In initially assessing a parcel of land, regardless of size, the various physical attributes must be carefully analyzed to determine how best to incorporate them into the master plan. These determinations should be made regardless of land use because a constraint in an office or industrial development might be an opportunity in a residential subdivision and vice versa. Examples of opportunities and constraints are provided in the following text.

Surrounding Land Uses

Surrounding land uses are perhaps the most difficult constraint to deal with because they are off site, or outside of the project boundary and not within the control of the project developer. This necessitates that any mitigation of potential negative impacts from the constraint be done on site, or within the boundaries of the project. For instance, an electrical power station or a landfill on adjacent property may be screened within the project by building an earthen berm and planting vegetation to help hide any unsightliness. Solutions such as this require allocating property that otherwise could be an income-producing element within the development. Some surrounding land uses are considered opportunities. Public parks, schools, or other community uses are generally considered opportunities, and plans for the project, particularly residential projects, should incorporate a connection of sorts (vehicular and/or pedestrian) to the complementary adjoining land use. Sometimes it is difficult to determine whether an adjoining land use will have a positive or negative influence on a new community. For instance, self-storage units are generally safe and well-lit at night, possess good security systems, and are quiet with little traffic—a great neighbor, in most instances. However, in some locations, self-storage units can be utilized for illegal purposes, such as drug manufacturing or warehousing stolen property. This usually introduces an unsavory element into or near the neighborhood—not a good neighbor to have. Grocery and service retail stores are generally considered positive surrounding land uses, whereas industrial sites, used car lots, and junkyards are almost always considered negative.

General Encumbrances

On-site encumbrances generally fall into two categories: physical characteristics and those that are imposed by a governmental authority as a condition of approval for the project. In either case, there is a net loss of developable property that must be considered when preparing a concept plan for the development project.

Physical Encumbrances

Some of the common physical encumbrances are easements for *electrical power transmission*, underground *pipelines* for the transmission of oil and natural gas, and other miscellaneous easements for access and utilities. In most instances, the crossing or other accommodation of these easements can be negotiated with the easement's owners, usually at some cost. In other instances, easements can be relocated and/or eliminated if they are no longer needed. The cost of this option is typically much higher and must be compared to the increased efficiency of the resulting plan.

Because pipelines are underground and extend long distances, they present a great opportunity to install trails and become an integral part of a community's overall pedestrian and bike trail network.

Electrical transmission lines are above ground and unsightly, but they are still linear and extend for long distances, creating a great opportunity to incorporate them into the community's trail system. Figure 8.1 illustrates examples of physical encumbrances.

Drill sites are parcels of land that are reserved for the potential drilling of minerals below the surface, such as oil and gas. It is common throughout the state of Texas for the owners of mineral rights below the property's surface to be different from those of the actual surface of the property. In those instances where this occurs, the rights of the mineral owners must be protected by designating specific areas in which future drilling could occur. These sites usually range in size from one to ten acres (with three acres being typical) and are located strategically to serve an area below the surface within a half-mile radius, although directional drilling can sometimes far exceed that distance. The size, shape, access, and location of these drill sites can have a significant impact on the land plan for a development. Some developers choose to shove the drill site into a corner, fence it, provide access to it, and be done with it, whereas others try to incorporate the sites into an open space plan or another temporary or removable use dependent on the likelihood that the surface will actually be utilized for drilling. A few interim uses for drill sites include open space, sports fields, playground equipment, and parking lots.

Rivers, Creeks, and Bayous

Natural water features across a tract of land can have an impact on the ultimate land plan from several perspectives. Not only do they typically serve as the primary drainage facility for a development project, but they also can be embellished with landscaping and trails to become an aesthetically pleasing and functional amenity for the community. Drainage facilities also can serve as natural barriers between land uses and separation of neighborhoods. These natural and/or improved features can become a central focal point within the community when the overall master concept plan acknowledges and accommodates their potential.

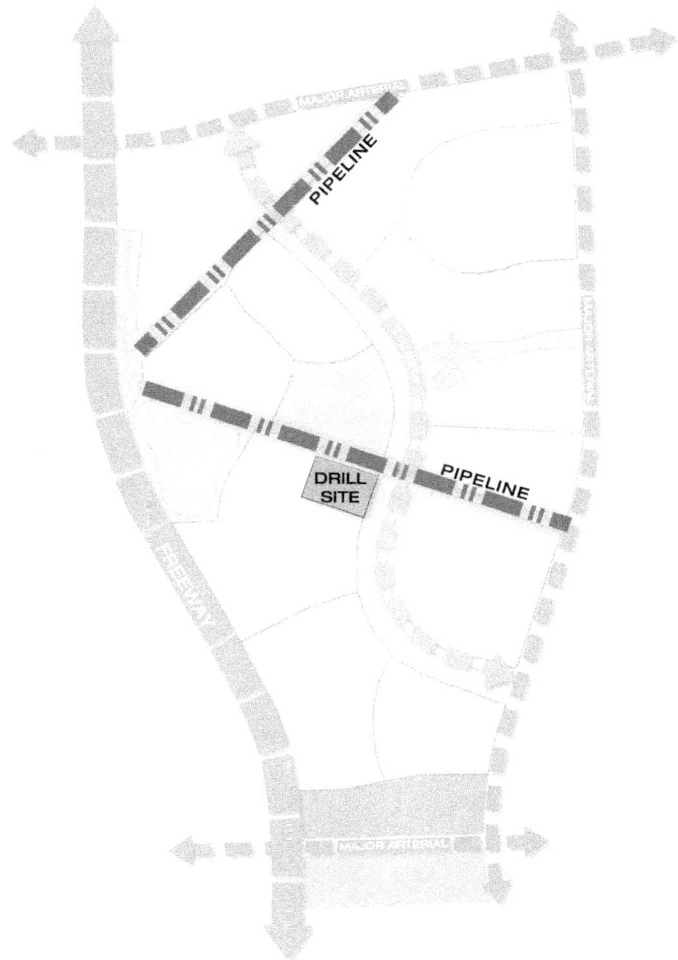

FIGURE 8.1 Physical encumbrances

Contours/Elevation/Topography

Depending on the severity of the topography in a region, elevations and contours can have a major influence on the land plan for a project. In hilly terrains, drainage corridors are naturally defined and not easily altered. The percent slope on specific parcels or lots may necessitate fill dirt being added or retaining walls being constructed, which affects the efficiency of the plan and the development. Driveway and sidewalk slopes must be considered for safety purposes. On the other hand, hilly terrains provide the opportunity for creating view corridors from public spaces and lots with careful orientation. In flatter regions, the lack of topography means that view corridors must be created by landscaping or hardscapes that include trees, monuments, water features, or buildings that are architecturally interesting. Drainage usually consists of manmade channels or underground pipes (storm sewers), providing some flexibility in location. Regardless, topography plays an important role in the overall design of a community.

Government-Imposed Encumbrances

Discussed below are typical government regulations that are not initially present as part of the physical characteristics of the property but designed and constructed as part of the criteria for obtaining approval of the development project.

Virtually all jurisdictions have an adopted *major thoroughfare plan*, which proposes a network of streets designed to facilitate future traffic demands. Most of these plans are implemented by developers, sometimes in cooperation with a public agency, on a project-by-project basis. As development occurs, the roadway network is connected and expands. Criteria for various categories of roadways typically vary from jurisdiction to jurisdiction in terms of right-of-way width, paving section, and curve radii, among other characteristics, but generally, there is enough flexibility to allow these roadways to be integrated with the proposed development plan. In instances in which a tract or parcel of land is affected by a proposed thoroughfare, the property developer must dedicate the land (i.e., the right-of-way) to the public for the roadway and, in most cases, construct the thoroughfare according to the agency's specifications.

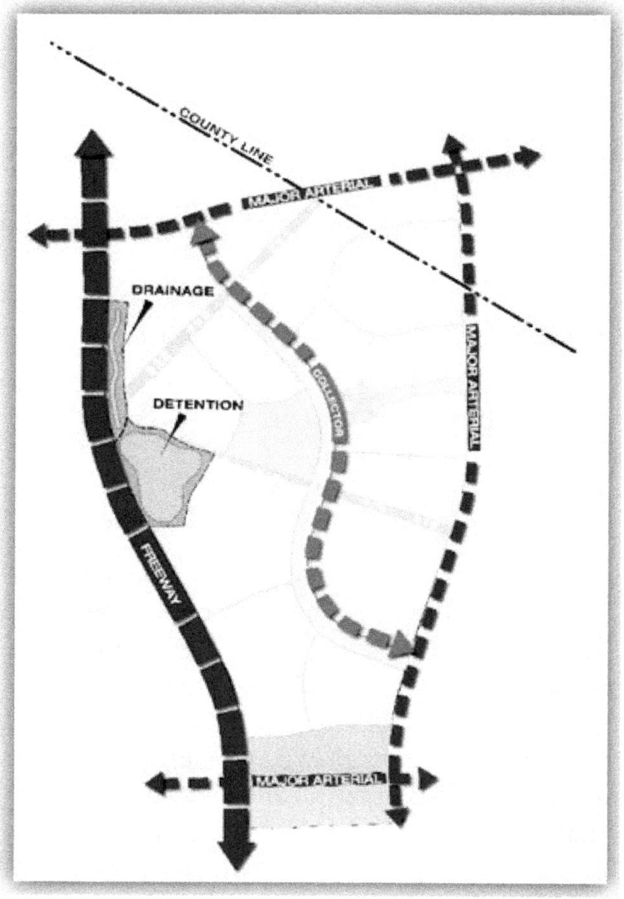

FIGURE 8.2 Government-imposed encumbrances

Detention ponds are a common government regulation designed to mitigate the impact of new development on surrounding properties, particularly those that are downstream. The idea is to retain and/or detain storm water for a period of time so that stormwater runoff is not released all at once into creeks and bayous that serve the general area and thereby avoid possible downstream flooding. Depending on the region's propensity for flooding, the amount of land required to satisfy this requirement can be significant. Figure 8.2 illustrates examples of government-imposed encumbrances.

Detention/Retention ponds generally fall into two categories: dry and wet. Dry ponds are exactly what the label implies—a large hole in the ground, typically six to ten feet deep, that remains dry until it collects stormwater runoff from a rainfall event. Then it is slowly released to a receiving downstream channel at a rate calculated by a hydrologist that is usually consistent with predevelopment flows. Dry ponds can be developed with athletic fields, walking trails, or other uses that can withstand occasional inundation for extended periods.

Wet ponds are a more prevalent method of handling detention or retention requirements. They are a much more aesthetically pleasing solution, particularly in residential communities and office parks, providing an attractive water feature that contains fountains and often stocked with fish. The wet ponds are obviously more expensive from the standpoint that they necessitate additional excavation to ensure adequate storage space for stormwater runoff, but there is a value-added element derived from premium lots that have water frontage and water views.

CONCEPT PLAN

These plans are sometimes referred to as *bubble plans* and typically illustrate proposed land uses, major roadway systems, lot projections, and various project features. Concept plans are dynamic and are constantly manipulated and updated throughout the development process. Simply stated, a concept plan is a puzzle with all the pieces representing the various constraints, the transportation network, and various land uses, including parks and open space. When the pieces of the puzzle are properly arranged and the various land uses have been appropriately situated to serve the community, the results can be far superior to those of most cities and towns with strict zoning restrictions. The ability to take a blank canvas and select the location for every land use allows for the ultimate plan to work with very few, if any, land use conflicts; it also permits traffic patterns to conform with area demands, decreasing congestion and distributing traffic onto a roadway network designed to manage it. In this way, a concept plan essentially serves as a zoning map that is dynamic and can be adjusted without regulatory bureaucracy. This private-sector solution works far better than the actual zoning restrictions in most cities. Figure 8.3 illustrates the various components in a concept plan.

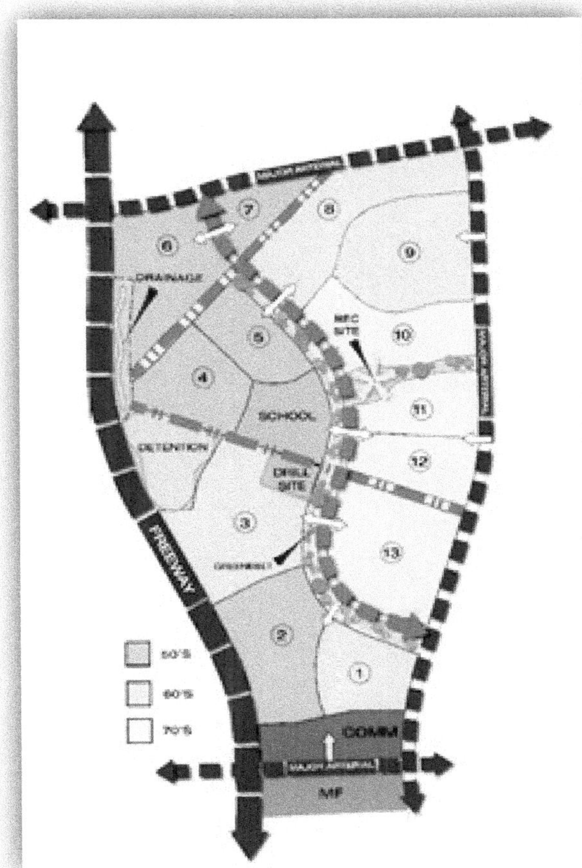

FIGURE 8.3 Concept plan

Future major thoroughfare plan (MTP): One of the first constraints to be addressed when preparing a concept plan is the primary street network. The MTP is not only the first step in satisfying a governmental requirement but also the first attempt at defining parcels for the master development plan. During this phase of the concept plan, planners are attempting to accommodate the various constraints within the property by strategically integrating the roadway network and shaping the parcels that will ultimately be utilized for various land purposes. The goal is to provide each parcel with clean or direct access to the primary street network (i.e., major thoroughfare and/or collector streets), providing the flexibility to alter land uses as market demands change. It is important to keep the sizes (and shape) of the various intended uses in mind during this part of the planning process. Some of the larger parcels will be further subdivided, whereas others will be permanently constrained with physical boundaries. It is critical to ensure that every parcel is capable of being utilized to its highest and best use, which means the planners must be familiar with the criteria—in particular the area size and dimensions—for the end users that are likely candidates for certain parcels.

Commercial: Commercial sites almost always prefer frontage on a major thoroughfare or collector street for access and exposure to high volumes of traffic. Even more preferable are commercial tracts located at the intersection of two major thoroughfares (see Figure 8.4). Commercial parcels vary greatly in size and dimension, depending on the end user. Pad sites for fast-food restaurants, banks, and other small standalone users require little street frontage and relatively shallow tract depths (typically one to two acres). On the other hand, large "box" users, such as grocery, department, and home improvement stores, require more frontage and deeper tract depth because of their large parking fields (usually larger than 10 acres). There is not a specific amount or percentage of commercial land that is allocated in a master planned community. In the past, developers assumed that all major thoroughfare frontage had commercial potential and would strip off the frontage for eventual commercial or multifamily use. This practice created an inordinate amount of commercial and multifamily inventory that was impractical and led eventually to undesirable land uses. Developers fell into this trap by not wanting to back lots and, ultimately, houses onto the thoroughfare because they felt they were not receiving any benefit from the roadway that they were required to dedicate and construct. However, as they realized how long they would need to hold the potential commercial property in inventory, it made more sense to develop some of the thoroughfare frontage as residential (backing not fronting) and, concentrate a more reasonable amount of

FIGURE 8.4 Commercial sites

commercial parcels near intersections, and extend commercial uses along the frontage only when warranted. Most retail stores base their location upon the area population or the number of residential rooftops. Dependency on rooftops is the reason that residential neighborhoods develop first before schools, retail development, and other uses. National chains, in particular, have extremely strict criteria, including site dimensions and area population, when it comes to siting new stores.

In recent years, a solution to the excess commercial property has developed that is beneficial to the public. Cities, counties, utility districts, and other special-purpose districts have begun purchasing these strips of property along the thoroughfares and developing them as linear parks, providing very visible and safe corridors for walking, jogging, and cycling. Figure 8.5 illustrates this solution and the contrast along the street frontage.

Civic/Community: It is important that civic and community uses such as schools, churches, libraries, and fire, police, and recreation facilities be strategically located throughout the community. Generally speaking, these uses should be located convenient to thoroughfares and collector streets, especially the emergency uses due to their need for quick access to the rest of the community. The siting of churches and libraries is not as critical as that of the emergency uses. Although they prefer to have similar access requirements, churches and libraries are more of a destination place in which travel distance is not the primary factor. Also, it's not unusual for churches of various denominations to be located adjacent to one another, so spacing or separation is not a requirement. It should be noted that the Texas Alcoholic Beverage Commission has enacted rules that restrict the sale and service of alcoholic beverages within certain distances of specific land uses, such

FIGURE 8.5 Strip commercial/park

as churches, schools, and day care centers, if the local jurisdiction chooses to adopt and enforce those rules and regulations. There are numerous exceptions and qualifications to the distance requirements, so caution must be taken to ensure that the latest requirements are consulted prior to making final determination of land use locations. Figure 8.6 illustrates the siting of the various civic and community uses within a large master planned community.

The siting of recreation facilities within the community is highly dependent on the size and use of the facility, and care must be taken to situate them accordingly. For instance, a small recreation facility with a neighborhood pool, a playground, and a couple of tennis courts functions nicely in the interior of a neighborhood and should be within walking distance of the area it serves. This type of facility does not tend to generate much traffic and is not disruptive to the neighborhood. On the other hand, larger recreation facilities with large pools and splash pads, large playground equipment, and multiple tennis courts usually serve multiple neighborhoods and attract a large number of vehicles for tennis tournaments, swim meets, and other community events. These uses should definitely be sited on the exterior of the neighborhood to minimize any negative impacts. An important associated use to all recreation facilities, regardless of location, is an integrated trail system. It is important to create connectivity throughout the neighborhood between community uses with a trail network that allows for pedestrian and bicycle access to each of the facilities. This should be kept in mind throughout the entire planning process and be an integral part of the overall master plan.

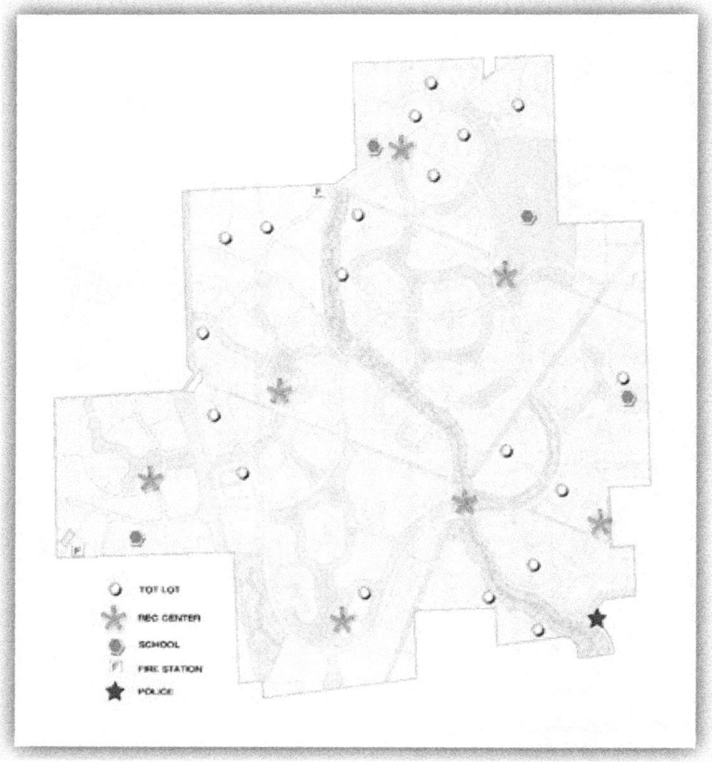

FIGURE 8.6 Civic/community uses

Schools tend to be one of the more difficult uses to situate within residential neighborhoods for several reasons. There is a natural tendency to place schools, particularly elementary schools, internal to residential neighborhoods; doing so allows children to conveniently walk or ride their bikes to school without encountering major traffic corridors. On the other hand, an internal location for a school means bus traffic must enter the neighborhood, sometimes in the early morning, creating a significant amount of noise and increased vehicular traffic (i.e., buses and cars). Most of the traffic is generated from outside the neighborhood where the school is located, which sometimes creates issues for the host neighborhood. Buses and cars are cued throughout the neighborhood several times a day, commingling with pedestrians and local neighborhood traffic. Peripheral sites on the edge of neighborhoods are sometimes better locations for schools, maintaining proximity and convenience while keeping the added traffic out of the neighborhood. Also, many school districts prefer the external locations to keep school sites close to the major traffic corridors. This saves on time as well as mileage and fuel for the multiple bus trips that are generated every day.

One last issue related to school site location is that occasionally a particular neighborhood may mature and stabilize without children, meaning that it does not have the housing turnover typical in most neighborhoods. More simply put, the children in the neighborhood grew up and went away while the parents stayed, leaving the school with few neighborhood students and forcing the school district to bring in students from remote locations. In some instances, the school site becomes so inefficient that the school district opts to sell the site as surplus property and abandon the site altogether. When this occurs, it is better for the school district to have a site that is located outside of a neighborhood and near a major traffic corridor. If the site is located interior to a neighborhood, the proposed new use can be a subject of major concern for the neighborhood. Likewise, junior high and high schools also present their own set of unique problems if not situated carefully relative to neighborhoods, due to their associated high-intensity activities and high traffic volume. Typically, these sites need to be placed along major thoroughfares. In recent years, school districts have opted to create large campuses for the placement of all levels of schools and stadiums, including bus maintenance facilities. This arrangement allows for the surrounding street network and on-site internal traffic circulation system to be designed to efficiently manage the large volume of traffic generated by school events and activities. Individual school sizes vary by district to district, but on average, an elementary school is typically located on 12 to 15 acres, a junior high or middle school is situated on 35 to 45 acres, and a high school is located on 75 to 125 acres. This includes ballfields and on-site parking.

Multifamily: An integral part of any residential community is the availability of multifamily units. Without delving into some of the social objections to multifamily units (or apartments) as a legitimate and necessary form of housing, a master planned community would not be complete without planning for and accommodation of rental housing options. Regardless of arguments to the contrary, newcomers with a need for temporary housing, minimum-wage earners, students, young people just starting their work careers, older adults, or those in need of age-restricted services need access to reasonably priced housing and are essential contributors to a community's economic viability. The strategic placement of multifamily units within a community is critical due to the importance of proximity to jobs, schools, and other community facilities. It is no less important to the multifamily complexes as it is to the single-family residential neighborhoods. Sites adjacent to commercial uses along major thoroughfares or adjacent to school complexes tend to be good locations for multifamily projects, whereas sites near single-family neighborhood entrances tend to be less desirable. Figure 8.7 illustrates an appropriate siting of a multifamily project within a community.

Single-Family: The designation of single-family parcels at the concept plan level serves several purposes. First, it demonstrates where the single-family parcels are to be in relation to everything else, including all the land uses mentioned above. Second, this designation allows the planners to quantify the single-family parcels and project a density for the overall project. This is done

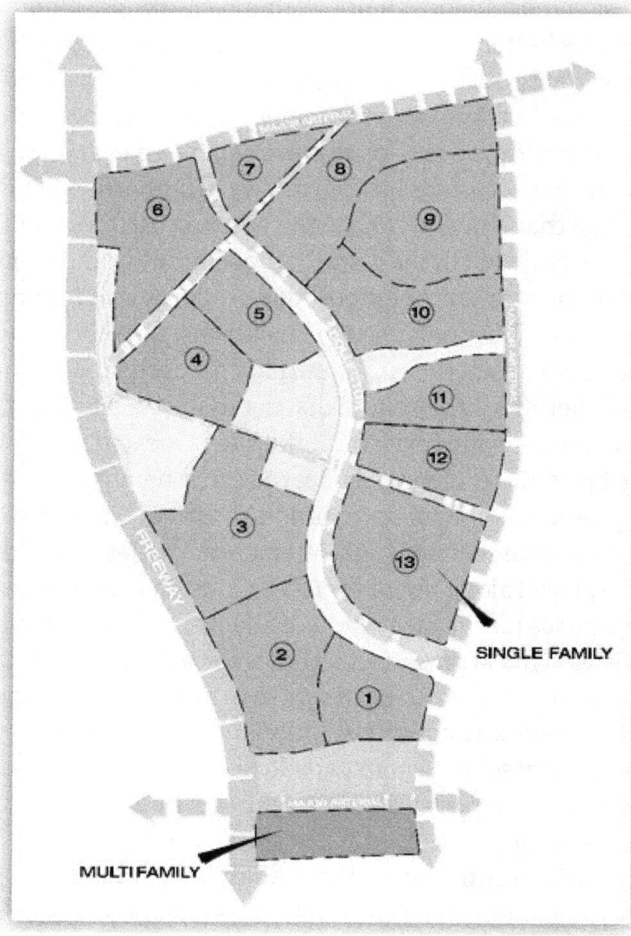

FIGURE 8.7 Single-family/multifamily uses

by calculating the acreage of each single-family parcel and then assigning a particular lot size to each. Because each particular lot size produces a specific lot density (i.e., lots per acre) on a given parcel, the typical density for each lot size is calculated for each parcel, providing a total number of lots for the project as well as a subtotal for each lot size. This allows the developer to approach potential builders for the project and determine their interest in the type and number of lots they may want to purchase. Typically, a master planned community will have multiple housing products (builders) and multiple price points so that the community contains housing products available for most potential homebuyers. This results in multiple lot sizes for the project. In theory, developers want to ensure they have a constant inventory of each price point available for sale in the community. For instance, if a particular community contains lots of 50-, 60-, 70-, and 80-feet width, that likely means that there are also four different price points for the home products. For the sake of this discussion, let's say the 50-foot-wide lots hold a house that sells for $250,000. That means the 60-foot-wide lot home will probably be somewhere in the $300,000 range, the 70-foot-wide lot will be in the $350,000 range, and the 80-foot-wide lot will be in the $400,000 range. Let's also assume that a lot that is 50 feet by 125 feet can be produced in a well-designed street pattern at a density of 4. 4 lots per acre, a 60-foot-by-125-foot lot has a density of 3.5 lots per acre, a 70-foot-by-125-foot lot has a density of 3.2 lots per acre, and an 80-foot-by-135-foot lot has a density of 2.6 lots per acre. This means that in residential cells containing 25 acres each, a planner could design a neighborhood containing 110 50-foot lots, 87 60-foot lots, 80 70-foot lots, and 65 80-foot lots, for a combined total of 342 lots.

Phasing

One of the more difficult tasks assigned to a land planner is ensuring that the concept plan can produce a constant inventory of multiple lot sizes to accommodate the various price points while simultaneously doing so in a phased approach, meaning that the first grouping of all sizes and price points must be in a confined area to ensure the efficiency of street and utility construction, including drainage. This means construction of the infrastructure cannot be scattered all over for the sake of separating products due to the increased expense of streets and utilities. So, the initial design of the residential cells within the concept plan must allow for a sequential phasing to occur, with each phase containing a full spectrum of lot sizes. That may or may not hold true for the more expensive homes because they typically do not sell at the same pace as those that are more affordable. Therefore, it is possible that a single phase could possess two actual phases of the higher-priced products.

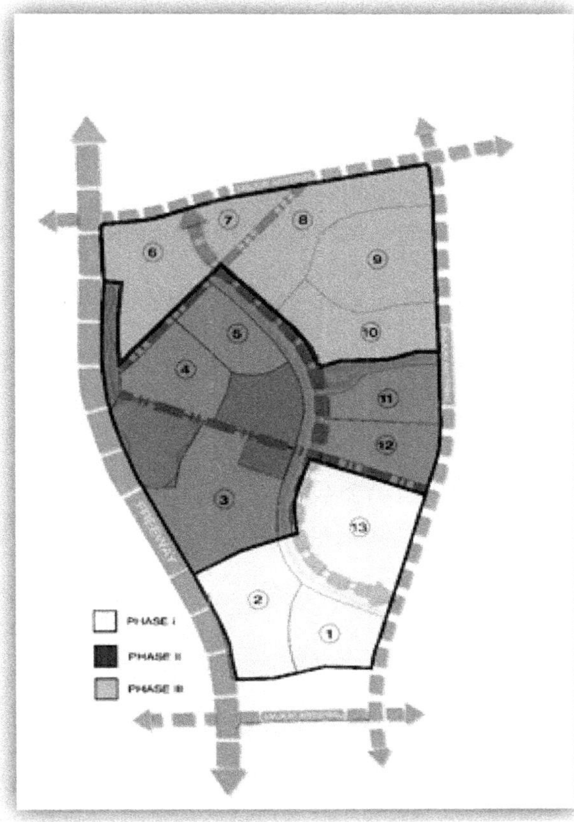

FIGURE 8.8 Phasing

Additionally, over time, the marketplace may change, necessitating a shift in lot sizes and requiring developers and builders to alter their plans accordingly. This is perhaps the most dynamic element of the concept plan. The individual acreage cells within the various phases allows for complete flexibility to respond to market conditions as far as lot sizes and price points are concerned. Figure 8.8 illustrates a phasing plan.

SCHEMATIC DESIGN

Street patterns tend to be one of the more permanent physical improvements in our cities and towns today, outlasting structures, property ownership, and land uses, particularly in suburban areas, where redevelopment occurs at a much less frequent pace than it does in urban areas. In 1997, Michael Southworth and Eran Ben-Joseph identified five distinct characteristics in the evolution of street design in our nation's cities. These are illustrated in their book, *Streets and the Shaping of Towns and Cities*, and range from the "gridiron" pattern prevalent in major urban areas to the "lollipops-on-a-stick" design that began appearing in suburban neighborhoods in the 1980s. And although advocates of new urbanism and neotraditional planning concepts still promote the perceived merits of the gridiron pattern, even in suburban areas, it seems it is a perception based more on nostalgia than on good planning.

In reality, what has occurred in our newer suburban master planned communities is another distinct characteristic in the evolution of street design. A hybrid design has evolved that incorporates both pedestrian and vehicular networks that are not only functional but also safe and secure. In a pamphlet prepared by Kerry R. Gilbert & Associates, Inc. (KGA Land Planning) in 2012, the firm examined the issue of connectivity as it relates to residential street design. Through high-quality, effective street design, KGA Land Planning demonstrated that it was possible to provide residents with a walkable community free of thru traffic without compromising public safety and emergency services. The result was quieter and safer neighborhoods, which pushed traffic onto collector streets and thoroughfares that were specifically designed to accommodate it. Dedicated greenbelts with a pedestrian trail system were incorporated into the neighborhood design, which permitted pedestrian trips to neighborhood amenities, including schools, parks, churches, and neighborhood retail services, without the conflict of local or thru traffic.

Schematic plans typically show proposed street patterns and preliminary lot layouts that serve as the basis for an engineer to determine water and sewer capacity as well as drainage and utility service plans for the project. A general indication of lot sizes and dimensions are illustrated in order to determine a more precise density for the project. This information helps the client determine projected costs for the project's overall development. Figure 8.9 illustrates a schematic plan.

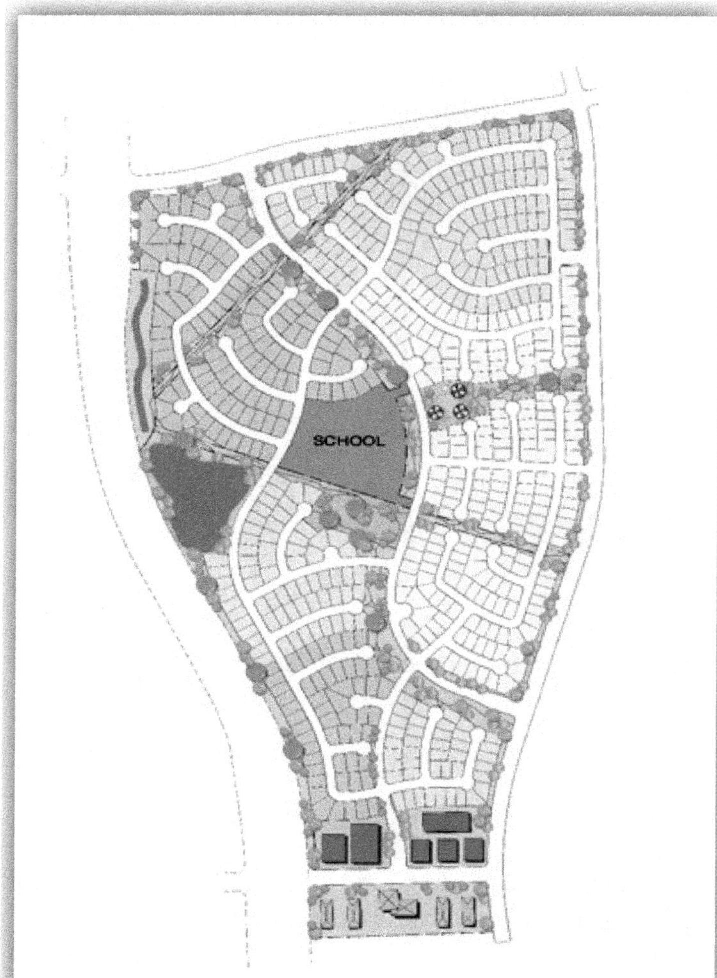

FIGURE 8.9 Schematic plan

Design Features

The design of the actual street patterns for a project can determine its ultimate success, especially from a homeowner's perspective. Although there are government regulations concerning the design of street and lot layouts, much has to do with personal preferences regarding what constitutes high-quality design. Numerous articles have been published regarding this issue, but in the end, it is the neighborhoods that prove to be most desirable by homebuyers that provide us with the best indication of what constitutes a good design.

Premium lots: Certain neighborhood lots are considered premium lots because they have frontage on a lake, greenbelt, or park. Other lots might be considered premium because they are oversized or situated on a cul-de-sac. The greater number of premium lots that can be created by the design of the street and lot pattern there are, the more desirable a neighborhood and its home sales velocity are. See Figure 8.10 for an example of premium lots.

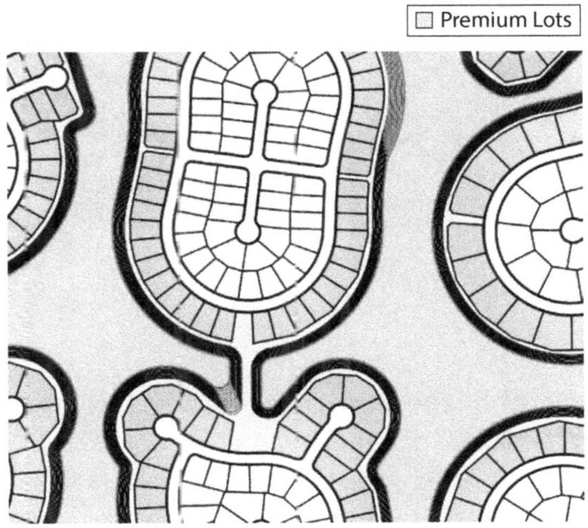

FIGURE 8.10 Premium lots

Cul-de-sacs: The utilization of cul-de-sacs in neighborhoods has proven to be an element that homebuyers consistently desire and are willing to pay for. Realtors consistently report that homebuyers are willing to pay a premium for a home on a cul-de-sac, and numerous studies have indicated that market values for homes on cul-de-sacs remain relatively higher than those not located on one. Residents generally cite safety, security, and a stronger relationship with their immediate neighbors as some of the advantages of living on a cul-de-sac.

Open cul-de-sacs: The open-ended cul-de-sac is a design feature that can be utilized in a variety of situations but is normally used in situations in which lots otherwise might be oriented to back onto a thoroughfare or collector street. Not only does this design feature not decrease the overall number of lots, but it also has several benefits for the neighborhood as whole. Rather than encircle the cul-de-sac with lots or extend a series of streets to intersect with a thoroughfare, the streets are terminated with a cul-de-sac and the end is left open, providing an opportunity to connect a pedestrian pathway from the cul-de-sac to the adjacent thoroughfare. This design solution opens the neighborhood visually but not physically for vehicular traffic. This result also avoids having a string of lots backing onto the thoroughfare, leaving the neighborhood walled in along the road. Because a wrought iron or tubular steel fence is usually used in this design feature, it saves on masonry walls that are typically installed along the thoroughfare as well. Figure 8.11 illustrates the open-end cul-de-sac concept.

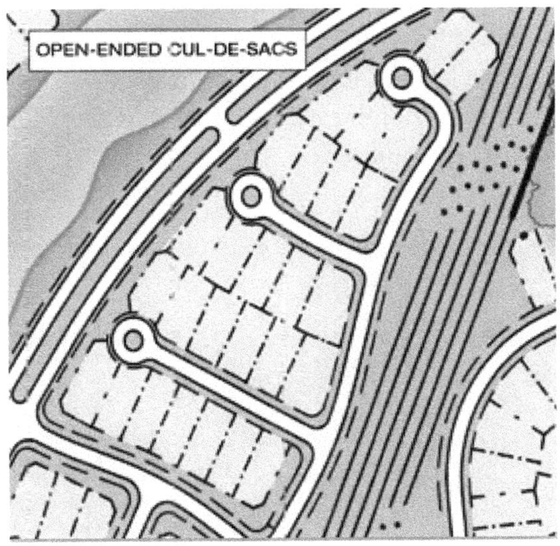

FIGURE 8.11 Open-ended cul-de-sac

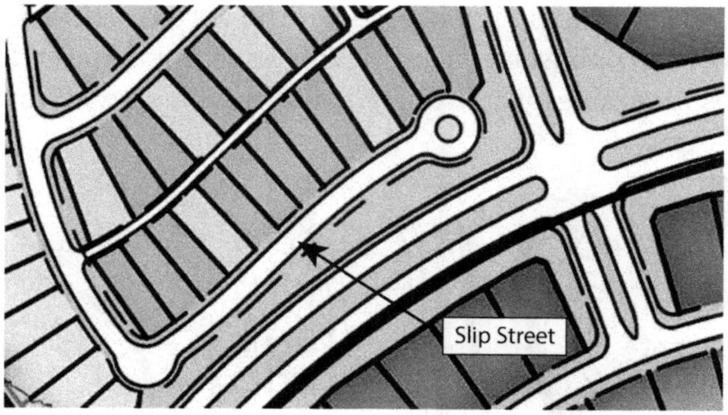

FIGURE 8.12 Slip street

Slip streets are those that parallel thoroughfares or collector streets, allowing driveway access to the lots that have frontage along these roads. This permits lots to have frontage along the thoroughfare without allowing driveway access. Albeit an expensive design solution due to the double paving involved, the slip street concept is an attractive alternative that enhances the aesthetics along thoroughfares and eliminates expensive masonry fences that typically line the thoroughfares where lots back onto them. See Figure 8.12.

Neighborhood entrances: An appropriately designed neighborhood entry can help define the character and identity of a neighborhood. For instance, a boulevard entrance allows for a neighborhood identifier in the esplanade and landscape reserves along the sides of the entry street, creating a grander look at the entrance to the neighborhood. With or without a boulevard, the entry street for a neighborhood can T into a park or lake at the first intersection, establishing a view corridor and giving the neighborhood instant character. This design element is far superior to a street serving as a primary entrance to a subdivision, which intersects the main thoroughfare and has homes that immediately have frontage (and driveways) near the intersection. This poor design feature usually encourages parked cars, which clutter the entry street, establishing an unsightly neighborhood entrance. See Figure 8.13 for an example of an appropriately designed neighborhood entry.

FIGURE 8.13 Neighborhood entry

View corridors: Internal view corridors can enhance both the aesthetics and value of the lots within a neighborhood. This is true not solely for premium lots but for all of the lots that are granted access to an amenity or a neighborhood feature because the feature has been exposed to the rest of the neighborhood via a view corridor intended to accomplish just that. In older subdivisions with lakes and parks, the developer attempted to put as many lots as possible around the amenity to maximize the number of premium lots but ended up effectively hiding the amenity and limiting the number of premium or value-added lots. Utilizing a street design that opens up or is exposed to a lake or park provides a view corridor that can be enjoyed by all the residents, not just those with frontage on the amenity. This design feature gives value to lots that do not have this type of frontage but are merely in proximity to the amenity. Lots that are located near greenbelts are similarly desirable in that homebuyers know that the greenbelts will take them directly to community amenities, and they are therefore willing to pay a relative premium for a lot within a certain distance of the feature. Therefore, an economic system relative to lot premiums is created, which assigns premiums not only to the usual amenity lots but also those that have view corridors and are in proximity to an accessible amenity. See Figure 8.14.

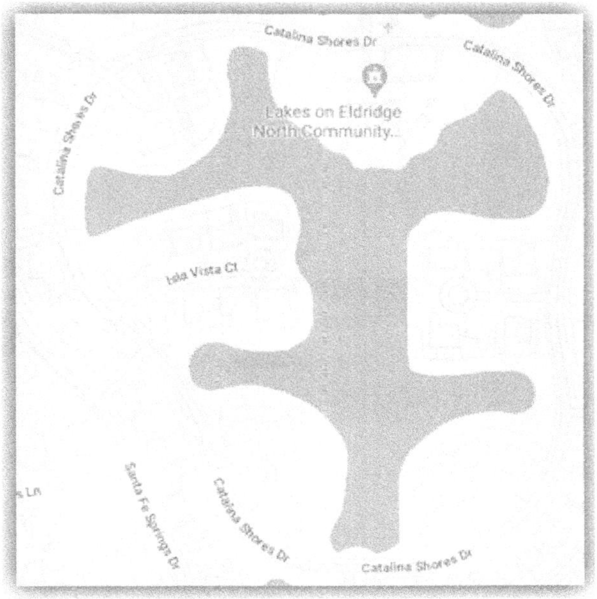

FIGURE 8.14 View corridors

Internal pocket parks and tot lots are another design feature that adds value, particularly for young families buying their first home. A well-designed plan can find and utilize inefficient parcels that lend themselves to small pocket parks and/or tot lots. This enables these features to be incorporated into the neighborhood plan without the loss of lots. Figure 8.15 shows a typical tot lot.

FIGURE 8.15 Tot lot

Gated subdivisions are not necessarily design features because the design is essentially the same as a neighborhood that is gated. Most communities in which gated sections are allowed require developers to construct their private streets to public street standards. The neighborhood then contributes to a maintenance fund that likely will not be needed for

FIGURE 8.16 Gated subdivisions

many years, and, in this way, the public bears no responsibility for the private streets' eventual upkeep. The maintenance of the gates and/or card key system also are typically paid for out of this fund. The amount each homeowner contributes to the fund varies with each neighborhood depending on the number of homes behind the gate and the expenses related to maintenance. For instance, some gated communities have staffed security gates, which adds significantly to their annual maintenance costs. Figure 8.16 displays a typical gated subdivision entry.

ENTITLEMENTS

Zoning

In the state of Texas, municipalities' ability to control land use is limited to their corporate limits, and therefore, most cities adopt zoning ordinances that designate a specific land use on specific parcels across the city. These required uses can be changed only after a public hearing is held to consider a change in land use.

Zoning cannot be imposed in the unincorporated areas of the state, including a city's extraterritorial jurisdiction (ETJ). Cities in Texas may impose development standards within their ETJ, typically through their subdivision ordinance, which may be extended into their ETJ. These regulations may include items such as street standards, building setback lines, and other technical requirements but may not address land use or development density.

Planned unit development (PUD): A PUD is a special zoning tool that allows for the creation of flexible zoning standards for a project when it is neither practical nor desirable to a have specific zones for certain parcels. The PUD tool allows for multiple land uses within an overall project boundary and, more importantly, provides the flexibility to shift and alter the proposed uses (within predetermined limitations) to better respond to project demands and market conditions.

Development agreement (DA): A DA may be used inside or outside of a city's corporate limits to achieve a mutual agreement regarding development expectations. The DA may address a variety of issues that include but are not limited to development regulations, utilities, reimbursements, land use, and phasing schedules.

Other tools utilized by a project's developer of that don't necessarily require interaction with the public agencies are as follows:

- **Deed restrictions:** These are imposed by a property owner on the overall development and generally restrict how property is utilized. These restrictions are recorded in the official property records of the respective county and are binding for anyone purchasing property within the development.
- **Residential builder guidelines:** These are documents prepared by the project developer to serve as an architectural control on the residential builders within the project and are designed to ensure that consistent quality is maintained from builder to builder and from phase to phase throughout the project. The guidelines are generally flexible to allow for market conditions.
- **Commercial builder guidelines:** These guidelines are included in a document prepared by the project developer to serve as an architectural control on the commercial builders within the project to ensure that consistent quality is maintained from commercial building to commercial building and from phase to phase throughout the project. These guidelines are generally flexible to allow for market conditions.

PLATTING

Subdivision regulations, from chapter 212 of the Texas Local Government Code, gives municipalities within the state of Texas the authority to promulgate rules regulating the subdivision of land within their corporate limits and extend those rules into their ETJs. Chapter 212 defines when a subdivision plat is required and describes the various types of plats as well as the specific application and submittal process for each plat type. Cities in Texas are granted a fair amount of leeway in establishing their own subdivision platting process as well as specific rules regulating development activity as long as they comply with chapter 212.

ETJs are authorized in chapter 42.021 of the Texas Local Government Code and are based upon population with smaller cities (fewer than 5,000 inhabitants) having an ETJ of 0.5 miles beyond their corporate limits and the largest cities (more than 100,000 inhabitants) having an ETJ of five miles beyond their corporate limits. Cities with populations in between those numbers are provided with incremental ETJs of 1, 2, and 3.5 miles depending on their population.

Preliminary Plats

Preliminary plats are required by most government agencies to determine the project's compliance, on a preliminary basis, with the adopted rules and regulations. Approval of the preliminary plat is basically authorization to proceed with the preparation of a final plat and construction plans for the project. Preliminary plats generally illustrate:

- Streets (dimensions)
- Lot dimensions
- Utility easements
- Pad sizes
- Building setback lines
- Additional characteristics; see Figure 8.17

Final Plats

Final plats are prepared by a professional engineer or surveyor and contain essentially the same data as preliminary plats but are much more detailed and have greater accuracy. Final plats are often accompanied by construction plans that illustrate the actual water lines, wastewater lines, and storm drainage systems as well as street paving plans. See Figure 8.18.

FIGURE 8.17 Preliminary plat

FIGURE 8.18 Final plat

VISION TO REALITY

More often than not, the process of planning a community from concept to reality requires the cooperation of a developer that has the same level of interest in creating a quality community as that of all the design consultants. It has been my experience that when you can demonstrate as a design professional that some of the design elements discussed above will not only contribute to the long-term value of a community but also facilitate sales velocity, most clients will listen and consider their options. Once the marketplace proves that a plan feature or element results in increased sales, then the competitive nature of the real estate development and homebuilding industry will dictate that those elements be repeated in other communities.

Homebuyers are the ultimate decision-makers when deciding which residential communities possess the design features and amenities that appeal to them. To many, safety features are a primary concern, whereas others want plenty of recreational opportunities. The quality of the school district serving the community as well as the proximity of schools are always near the top of the list in terms of homebuyer preferences. The top choice among homebuyers for neighborhood recreational amenities has historically been access to a quality trail system, both local and regional. Figures 8.19 and 8.20 illustrate the transformation from vision to reality.

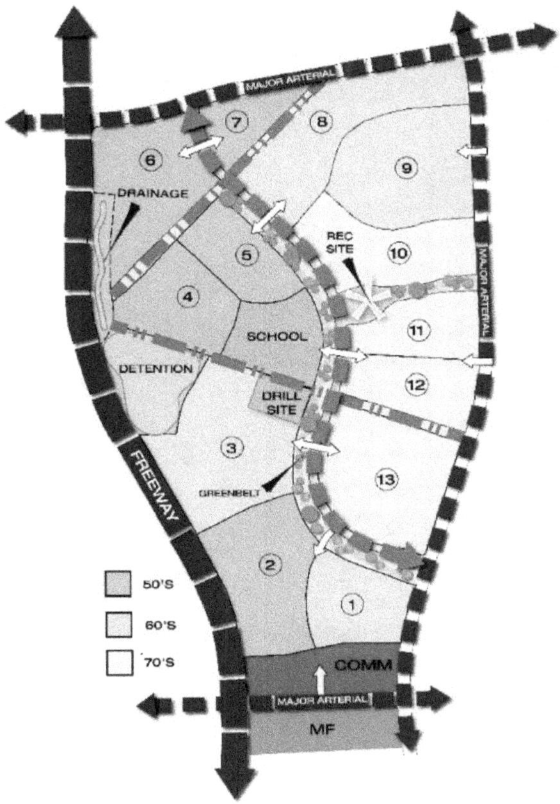

FIGURE 8.19 Concept plan

FIGURE 8.20 Project aerial

I have been fortunate to have designed numerous master planned communities throughout my career and also to live long enough to see how sustainable those communities have been. It's a track record I am proud of.

> "Excellence in land planning begins with a vision." —Kerry R. Gilbert

The image collage in the figures below is presented as an illustration of the various design components of the completed master plans.

FIGURE 8.21A–H Collage of completed master plans

FIGURE 8.21A Shadow Creek Ranch, Pearland, Texas. This image illustrates the design of the drainage detention system as a lake amenity throughout the community, producing numerous premium lakefront lots. This photo is a good example of the commingling of a technical function with an aesthetic feature and providing a functional and usable amenity within the community.

FIGURE 8.21B Bridgeland, Harris County, Texas. This image illustrates primary neighborhood entries from a collector street within the community and defining individual neighborhood cells. This specific design allows for a long linear park along the collector street, giving a visual break from having houses backing onto the collector street. The use of roundabouts and traffic circles serve as a traffic control element, reducing speeds and increasing safety within the community. This photo illustrates the traffic circle situated immediately inside the residential neighborhood, marking the entry and identifying the specific neighborhood.

FIGURE 8.21C Shadow Creek Ranch, Pearland, Texas. This is another image that illustrates a boulevard entry into a well-defined neighborhood cell, which provides an opportunity create an upscale entrance into the neighborhood and distinguish it from others in the community. This aerial oblique shows the landscaped bridged entry penetrating the neighborhood from a dedicated collector street, giving the secluded neighborhood an entrance unlike others in the community.

FIGURE 8.21D Cypress Creek Lakes, Harris County, Texas This image illustrates the continuity of the lake amenity system throughout the community and how it is exposed to the collector street system that serves the various neighborhoods at strategic locations. Also note that the pedestrian trail system surrounding the lake amenity system is connected at this specific location, allowing for pedestrian connectivity throughout the community.

FIGURE 8.21E Aerial photo of Cross Creek Ranch. This image demonstrates a unique creation of an amenity incorporating a major drainage facility within the community. A creek has been expanded and manipulated to become a meandering stream, one winding its way through the neighborhoods and creating a large, linear amenity that incorporates pedestrian trails, wildlife, fishing, and other aesthetic attributes. This photo depicts a collector street with pedestrian pathways and lighting parallel to the creek with periodic bridges crossing the creek to provide vehicular access to adjoining neighborhoods. This photo illustrates the magnitude of this type of project, the impact of which on the overall community is clear to see.

FIGURE8.21F Riverstone, Sugar Land/Missouri City, Texas. This image illustrates another example of utilizing the connective attributes of the overall drainage system as an amenity for the entire community. From pedestrian trails to major boulevards, connectivity to individual residential neighborhoods and recreation facilities is an integral element of the overall community design.

FIGURE 8.21G Pomona, Brazoria County, Texas. This image illustrates the siting of a recreation and playground facility on the periphery of a neighborhood yet easily accessible via a comprehensive pedestrian trail system. The facility is highly visible from the adjoining thoroughfare and collector street system, which incorporate crosswalks for pedestrian safety.

FIGURE 8.21H Grand Lakes, Fort Bend County, Texas. This image illustrates the benefits of an internal street system designed to create defined neighborhoods while allowing for an integral pedestrian trail system and multiple water features. The use of cul-de-sacs, traffic circles, and nonloaded collector streets allows for the safe and efficient movement of traffic while promoting connectivity of the various amenities throughout the community via a comprehensive trail system.

REVIEW QUESTIONS

1. Land planning starts with an examination of the property's opportunities and constraints, each of which must be analyzed and factored into the project plan. Why is surrounding land use one of the most difficult constraints to plan?
2. General encumbrances fall into two categories: physical encumbrances and government-imposed encumbrances. What are examples of physical encumbrances, and how might they be managed?
3. What is the purpose of planning a drill site inside a residential community?
4. Rivers, creeks, and bayous create some great opportunities, both functional and aesthetic. What are those benefits?
5. What impact does topography have on drainage, and what consideration must be given to access to the lots?
6. What is the concept plan, and what purpose does it serve in designing the project plan?
7. Which is the MTP, and how is it the first step toward satisfying a governmental requirement and the first attempt at defining the parcels for the project?
8. Describe the parcel size and depth issues regarding commercial parcels.
9. When it comes to civic and community uses, what role does the trail system play in creating connectivity throughout the community?
10. Describe some of the considerations given when locating a school within a project.
11. Why do single-family parcels have multiple sizes of lots?
12. What is the goal of project phasing?
13. The schematic design is a more precise project plan that defines street patterns and project density. How will this plan be used by developers when approaching prospective builders?
14. What are the benefits and challenges associated with various street design features, such as premium lots, cul-de-sac lots, and slip streets?
15. When developing a gated community, what considerations must be given to the streets?
16. What are entitlements, and how must they be considered when land planning a community?
17. What is a plat, and what steps does it take to become a final plat?

REFERENCES

Ewing, Reid. *Best Development Practices: Doing the Right Thing and Making Money at the Same Time*. Oxfordshire, UK: Routledge, 1996.

Kerry R. Gilbert & Associates, Inc. *Neighborhood Hybrid Design*. 2012.

Southworth, Michael, and Eran Ben-Joseph. *Streets and the Shaping of Towns and Cities*. New York: McGraw-Hill, 1997.

Texas Local Government Code, chapters 212 and 42.021.

Figure Credits

Fig. 8.16: Copyright © 2012 Depositphotos/tomfawls.

Fig. 8.18: Adapted from: LJA Engineering & Surveying, Inc., Final Plat, Shadow Creek Ranch, Pearland, Texas, 20 Apr 2004; adapted by Kerry R. Gilbert and K. Michelle Stoops.

Fig. 8.20: Copyright © 2021 by Google.

Fig. 8.21a: Copyright © 2019 by Shadow Creek Ranch. Reprintec with permission.

Fig. 8.21b: Copyright © 2020 by Smiley N. Pool/Houston Chronicle.

Fig. 8.21c: Copyright © by Shadow Creek Ranch. Reprinted with permission.

Fig. 8.21d: Copyright © by Mischer Investments. Reprinted with permission.

Fig. 8.21e: Copyright © by Cross Creek Ranch.

Fig. 8.21f: Copyright © 2019 by Johnson Development Corporation. Reprinted with permission.

Fig. 8.21g: Copyright © by Hillwood Development Company, LLC. Reprinted with permission.

Fig. 8.21h: Copyright © 2021 by Google.

THE ARCHITECTURAL GROUP

CHAPTER 9

INTRODUCTION

The architectural group is in many ways the soul of the company. This group's work product is the creative and technical manifestation of the company's product vision.

Large production builders all have their own in-house architectural group for plan creative design and plan document control purposes. Initial plan designs may originate from in-house designers or outside architectural groups, but these large-volume production builders produce almost all their construction documents with their own employees inside their own plan shop. Construction Science and Management (CSM) graduates who go on to become homebuilders will get to play architect, as all homebuilders eventually active their inner-architect self in the designs of their home products.

WHY IN-HOUSE ARCHITECTURAL?

Large production builders need to completely control their own home products and the architectural process and schedule, from the plan conception to the final construction drawings.

The Home Product Defines the Company

Production builders control the architecture to control their own market identity. Homebuilders are what they build. The home product is the manifestation of the company to the market. The art of homebuilding is in the composition and design of the home. All homebuilders think they produce better home designs than those of their competition. Why would they ever design a home they thought was a poorer design than the competition? The home is how each builder presents themselves and competes in the marketplace. Builders want to personally express themselves in their homes, even at the corporate level of a big company. Because better-designed homes sell more quickly and for more money, the home product design is at the center of a builder's likely success.

Plan Document Consistency

Production builders need consistency in their plan documents. Drawings of the same exact floor plan created by different architects will result in plan sets that are significantly different in appearance and in how they are read. Different architects use different symbols for standard items as well as different plan layouts, nomenclature, symbols, formatting, and plan protocols when they produce their construction plan sets. Production builders want all their plans to look and read the same so that their employee builders can automatically read them and know all the symbols and where to find all information they need. Plan consistency is one component of the system that supports the production builder's ability to build many homes simultaneously.

Control the Process and Priorities

Production builders want control over the plan shop's work priorities. Home builders change their priorities often. They need to be able to tell the plan shop on Monday that the priority is plan X and then on Wednesday that the priority is now plan Y. Production builders can drive an outside architect crazy with all their shifting priorities. The plan shop must create the plans needed to build the homes on the month dictated by the annual business plan. These internal goals are best managed internally as market conditions can shift and change a company's priorities very quickly.

Create the Team

Production builders need a lot of architectural capacity. In fact, in this professor's more than 37 years in the business, the plan shop was never once caught up; there was always a backlog of work. When this backlog occurs internally, the builder can respond internally and hire more computer-assisted design (CAD) operators and gear up and down to suit their business plan. This need for greater architectural capacity has stretched the internal plan shops to such an extent that there have been cases in which production builders have bought entire architectural firms just to capture 100 percent of those firms' design talent and plan production capacity.

THE ARCHITECTURAL GROUP ORGANIZATION

There are two functions performed by the architectural group: creative home design and plan document management, including the initial production and ongoing plan maintenance.

The architectural department leader is usually the licensed architect on staff who signs and seals all the construction documents for the municipalities. However, being a great home designer does not automatically make someone a great architectural department manager. In fact, these are essentially two different skill sets all together.

Exceptionally large builders often have a vice president of product design who is the creative product leader for the company. These builders may also have a position for someone whose sole responsibility is running the operation of the plan shop, normally called the plan shop manager. That administrative and supervisor position manages the CAD operators and plan document production process. This role is the employee responsible for managing the plan production group's ever-shifting priorities. The plan shop manager typically does not have creative design responsibilities, only plan shop production responsibilities.

Many national builders have nationally centralized plan product control. For example, a builder may build in Texas; however, their plan design and documents come out of California. Other national builders allow their local divisions to make their own plan product decisions and manage their own plan documents locally. There is efficiency and control in managing plans centrally, but it can lead to a loss of local architectural flavor as well as the local city/division managers losing document control over their plans. I always had direct control over my designs and documents, and I would not want to do it any other way.

The Two Duties of the Architectural Group

The architectural group carries a great deal of responsibility as the group's plans have a significant impact on the success of the homebuilder. Great companies have great plans.

Creative Origination of New Plan Products

The architectural group either (1) self-designs new plans, (2) creates new plans from existing popular plans, or (3) contracts for creative plan design services from an outside house design firm.

The architectural department manager must anticipate the company's architectural needs based on the new projects included in the annual plan. The director of product design must act far enough in advance to have new designs ready to meet the start dates contained in that annual business plan. For example, if the company plans to build a new patio home community starting late in the upcoming year, the design director must anticipate from the annual plan the company's product needs and determine what will become their plan lineup for that new community. Does the company's plan lineup already have the plans they need? If not the time to start designing is as soon as possible to be ready with the plans as needed to hit the critical start dates and accomplish the goals of the company's business plan.

Plan Production and Maintenance

The construction document production process is usually controlled by a plan shop manager. This plan production work will consume almost all the shop's CAD capacity. The other side of running a plan shop is plan document maintenance process.

Plans are always changing due to either an issue found in the permitting process or when building the home or changes requested by the sales team aimed at improving the plans. Plan document maintenance is an ongoing process for which the work demand is always underestimated. This type of maintenance is a separate category of work from plan design or plan production, and it is worth mentioning that plan changes will significantly affect the workload of a plan shop.

Architectural Shop Internal Collaboration

The architectural group is constantly collaborating with all other company departments. In some ways, this group stands directly in the center of all the back-office corporate operations.

- **The plan group collaborates with the land/lot acquisition group** in planning new products for new communities. Developers want to know what prospective lot buyers are going to build, and the architectural group will produce the product design exhibits needed to cast the builder and their products in the best possible light.
- **The plan group collaborates with the purchasing group** to estimate the cost of new plans and new plan details and new features under consideration during the design development process. Early drawings of new plans are sent to the purchasing group, which will calculate a construction cost estimate from a partially complete plan set, known as a *design development plan set*. This early estimate gives the plan design group early information about how the house might be revised to hit the targeted retail price point. Some new plans look great but may push the house cost out of the targeted price bracket.
- **The architectural group collaborates with the construction group** regarding the buildability and repeatability of plan designs and features. Builders often say the house is easier drawn than it is built. Just because they can draw it does not mean it can be built that way.

Production builders do not build just one of these houses; rather, they need to be able to build that house repeatedly, and each next version of that house must look and be constructed the same. It is like food at a restaurant, for which there is an expectation that it will consistently prepare a certain dish in a specific way. It is the same in production homebuilding—the plans must be designed to be built easily, using conventionally available materials and methods so that the floor plan and architectural features can be built the same way every time.

The plan shop group and builder will walk the new houses after the frame or drywall stage to critique the house as designed and built and review issues discovered by the builder during the construction. An example of a change that might be identified in a frame walk would be the unplanned—but necessary—placement of an HVAC (heating, ventilation, and air conditioning) chase inside one of the closets. That new chase need becomes a revision to the construction and marketing drawings that the buyers will see on the website or in the sales office. The plan is changed to reflect all the changes found during the frame walk.

CONSISTENCY IN THE PLAN SHOP

Production builders need consistency in their format of their plans—not just for the builders reading the plans, but also to facilitate productivity inside the plan shop. When CAD operators know exactly how they are supposed to produce a plan set, production moves much more quickly. If they did have a question, they can simply review another plan set and display that condition in the same manner on the new plan. Because everyone is using the same plan set completion checklist for every plan, over time, plan development becomes easier and faster.

Plan Master Sets

The completed master set of drawings are maintained by the plan shop to be current to all approved changes. The dated master plan must match all the material takeoffs, subcontractor budgets in the purchasing system and that dated plan version is the plan from which all sales prices apply.

Those master sets of drawings are made available electronically to other departments and approved subcontractors and vendors. Today, electronically viewing plans is probably more common than doing so via paper. Electronical access by employees to plans is almost universal inside the company. Builders always have access to plan sets electronically in their office or on their mobile device. Construction partners, vendors, and trades are granted plan access based on their needs, and if a subcontracting relationship ends their plan, access can be turned off. Electronic access ensures that everyone is always looking at the same dated version of the plans, because older versions are not displayed online. Each house will only show that house's specific plan set which matches the master plan as of the start date and any plan options selected by the buyer. The electronic system will only let the builder and subcontractors see the house's correct plan version.

Job Plan Sets

The Plot Plan

The plot plan process may be managed by the starts group or architectural group, but ultimately the plot plan will become part of the complete plan set, which is managed by the architectural group. The plot plan is likely produced by the surveyor as part of that company's scope of work and is used by that surveyor for the form survey, final grade survey, and final survey.

Job-Specific Architectural and Engineering Sets

The job-specific architectural plan set of drawings for each house started is produced internally by the plan shop. On a spec house, this could be as easy as printing the current base house prints. That ability to start houses efficiently and quickly is clearly one of the advantages of production-style building. On the other hand, the plan set for a sold home may need various pre-drawn and pre-estimated options added to the plan set. The system can add these approved options relatively easily, and usually there is a dedicated CAD operator doing nothing but revising plan sets to precisely match each upcoming sold start.

Once those job-specific plans are complete, they are physically or electronically delivered to the construction office as part of a starts package. Inexpensive, high-quality printers in the construction offices allow the builders to produce their own paper plan set in 1/8-inch scale on legal-sized paper when needed.

SUMMARY

The architectural group controls the company's product and market identity. CSM graduates who become homebuilder executives will be responsible for the company's product and architectural resources. Not only is the architectural group a big operation, but it is also sitting in the center of company's entire market value proposition and competitive strategy. The group is usually led by a licensed architect but may also be managed by an operating manager, the plan shop manager, who focuses on the production and not on the creative side. New plan designs originate both internally and externally but are always managed from inside the department by a licensed architect acting as the company's product director.

REVIEW QUESTIONS

1. What are the four main reasons production builders produce all their architectural drawings using their own employees in their own fully operational plan shop?
2. Who usually manages the plan shop, and who is most often responsible for the creative side of the plan shop? Who manages the plan document production?
3. Why is the creative design side of the plan shop a licensed architect?
4. New plan production and existing plan maintenance are the two basic duties of the architectural group. How are these tasks different, and how are they managed?
5. With which other departments does the architectural group routinely collaborate, and on what specific topics do they usually work together?
6. Describe the concept and application of the master plan set of drawings for each active floor plan and the job-specific plan sets. How do production builders produce a job-specific plan set when buyers select various options, and how are spec homes' options and selections made?
7. How do builders use electronic plan access to manage plan documents for bidding and construction?

THE PURCHASING DEPARTMENT

CHAPTER 10

INTRODUCTION

Construction Science and Management (CSM) students should note that the existence of a centralized procurement department inside homebuilding companies is an important distinction of homebuilders and the commercial construction company project teams who perform the buy-out of each project at the project level.

In homebuilding companies, the construction procurement and supply chain management function has evolved into a highly sophisticated corporate office department. This department consists of a team of procurement specialists who centrally perform the bidding and awarding for all the materials and most of the labor subcontracts across all the company's community projects and for all the homes they build. The production builder participates in the procurement process, but that process is led by the purchasing department.

A Great Place to Start a Career

I should confess that this professor had the privilege of starting my homebuilding career in the purchasing department at David Weekley Homes. I spent my initial 12 years learning the industry from the inside a couple of companies' purchasing departments.

I recommend purchasing as a career for many CSM students. If you don't choose this as a career, I recommend a few years working inside the purchasing department. Professional homebuilders who have experience in both the construction and the purchasing groups are in high demand and are well prepared for promotions and additional responsibilities and for an exciting and rewarding career in the homebuilding industry.

The following are five reasons to spend a few years in a purchasing department:

- **You will learn about every component** used in the construction of a new home. Further, you will learn the advantages, disadvantages, and true net costs of each alterative. The purchasing department investigates and approves every component that becomes part of the new home. Selecting the best components takes expertise and a lot of study. New products are constantly being presented to the purchasing group by manufacturers and supplies. The purchasing group must evaluate each new method and component.

- **You will learn all the different methods of installation and home construction** and about new and innovative systems that might be used in building the homes. You will learn the advantages and disadvantages of each method and system. Contractors often have different methods of performing the same task, and it is the purchasing department's responsibility to determine which best meets the company's goals and then write that method into the company's specifications and the subcontractor's scope of work.
- **You will learn the total house cost formula.** You will become an expert at precisely writing the construction specifications to intentionally move construction costs up or down. You will learn which specifications are most impactful to total cost and how you can obtain more perceived value with lower actual costs. You will learn why one home cost more to build than another. And you will be able to apply this knowledge during the rest of your career.
- **You will learn the homebuilding back-office methods of operation.** Most homebuilding companies have half their employees in the field and the other half in the office. Understanding the corporate office half of the business is just as important as learning the operations side in the field. As the CSM graduate's career matures, both the field and office tasks will someday become their total responsibility. Learning the corporate office operations side of your company is extremely important.
- **You will be discovered by the executives in your company**—not only as a builder, but as a homebuilding contract and cost expert. The purchasing group is a high-profile group inside all production homebuilding companies. Their cost projections so profoundly impact the company's profitability that the purchasing manager is often a highly senior member of the executive team. Your outstanding work inside a successful purchasing group will be highly visible, and from that exposure, new opportunities will open during your career.

THE EVOLUTION OF "PURCHASING"

The task of estimation originated from the employee who was once the builder's skilled assistant, performing material and area quantities takeoffs, and evolved into a builder's representative who negotiated and awarded the construction subcontracts. As the estimators' technical skills grew, so did their responsibility. Eventually, the estimators were handed the entire responsibility of negotiating and awarding the subcontracts and completing the house construction budget on the builder's behalf. The purchasing department evolved from a simple sole assistant into a sophisticated procurement department. Today, the purchasing function has expanded to an even higher level of sophistication, one responsible for managing a national and international supply chain of building materials and skilled subcontractors.

This evolution process is important to the CSM student because purchasing is an essential team inside all successful homebuilders, and therefore the purchasing department offers an important career opportunity for students entering the industry.

As homebuilding companies grew, the role and responsibility of estimating and contracting hundreds of millions of dollars evolved into a large department of construction specialists researching, evaluating, sourcing, recommending, and then contracting for new and better ways to build houses. The purchasing department became the company's research and development arm for the future components and methods that might be used in homebuilding.

CONSTRUCTION AND PURCHASING COLLABORATION

The most successful and well-respected companies have a healthy relationship between the purchasing department and the builders in the field. The issues the two groups must manage are challenging and can create stress between the two. The best company cultures create an environment in which these groups work collaboratively to create accurate material takeoffs and subcontract budgets.

CSM graduates who become homebuilders should strive to be great teammates with the purchasing department. Whenever you get a chance to enter their office, make a point to meet in person all the people with whom you talk by phone. Learn the purchasing department's responsibilities so that you will better know and understand your responsibilities and appreciate the great support services the purchasing department provides to you as a production builder.

THE PURCHASING DEPARTMENT'S BASIC DUTIES

The basic purchasing functions of this department include the initial bidding, bid evaluation, and subsequent awarding of construction subcontracts and material purchase orders (PO) to qualified material vendors, crafts, and mechanical trades whose prices are acceptable and whose scopes of work match the company's specifications. Even these fundamental functions are highly complex and important!

The basic purchasing functions of bidding and awarding construction subcontracts occur inside the smallest builders, just as they do inside the largest production builders. For CSM students who see themselves as small homebuilders someday, keep in mind that all these purchasing duties fall directly onto the small-builder category because each of the purchasing departmental responsibilities discussed herein must be performed to build even just one house.

Sourcing new subcontractors is a shared responsibility of the builders and purchasing department. Qualifying and onboarding new subcontractors and requalifying existing subcontractors is the responsibility of the purchasing department. The goal is to always have a ready pool of qualified and preapproved subcontractors from which the company's builders can choose and then hire to work in their community on their homes.

Introducing New Subcontractors

Production builders may introduce and start using any new subcontractor whom they wish to bring into the company's pool, but first there is a formal and important process to confirm that the subcontractor has the needed insurance, good references, and a solid material credit line if that subcontractor will be supplying materials. Additionally, there must be confirmation that the subcontractor's bid conforms to the company's written scope of work and specified components.

If a builder wants to introduce a new subcontractor, purchasing will evaluate that contractor's price against what the company is paying systemwide. The purchasing department is the group with visibility regarding the costs of every subcontract category across all the company's communities. If the price is lower than or equal to the company average, the department will start the process to onboard the subcontractor. If the subcontractor's proposal is higher than the all-community average, then the decision to use that subcontractor at the higher cost or renegotiate the price will likely fall to the project manager for that community. Paying more than the company average may be acceptable if doing so is the most efficient or only way to get the work done when and as needed.

Subcontractor Pool Maintenance

It takes a lot of work to continuously maintain the company's pool of prequalified, work-ready subcontractors. Purchasing usually performs the ongoing work to maintain current insurance certificates, safety plans, and whatever else is required to ensure these subcontractors conform to company policy and stay current in the company's payment system. The work of qualifying and requalifying the subcontractor pool is an important support service that the purchasing department provides to the builders.

The Subcontract Agreement

CSM students must understand another important subcontracting practice that is unique to production homebuilders: the **non-job-specific subcontract agreement**. Production builders utilize a **floor plan–specific subcontract agreement**. Note that custom builders use address-specific subcontracts because every home they build is a one-time build, unlike production builders, who build the same house plan multiple times.

Production homebuilders offer home buyers a choice of 10 or more floor plans in each community. Because the builder know that these are the plans that will be built in the community, the purchasing department will bid all those floor plans out to the subcontractors in advance. Once the plans' bids are agreed upon, the builder and subcontractor sign a subcontract agreement for each plan, and those bid prices become the budgets for each individual house plan, whenever it is built.

Additionally, here is the important point: The subcontract agreement does not guarantee that any or all these plans will ever be built, but if a given plan is built and that subcontractor does the work, then the agreement is already in place, and the subcontractor is ready to work at the plan's agreed-upon price and specifications.

When a house start is released, the builder contacts the subcontractor and schedules the work, knowing that the price and all the specifications were already agreed upon and signed into a legal agreement. The subcontractor performs the work and invoices at the subcontract agreement price, and the invoice is paid at the subcontract agreement price without a cost variance. The process works very well.

The production builder's subcontract agreement is not binding for either party. This is important! If the subcontractor finds other work or is somehow no longer interested in working for the builder in the future, the agreement is not binding for the subcontractor to work. Alternatively, if the builder either never builds those homes or uses another subcontractor for the work those homes, the agreement is not binding for that party either.

In summary, the floor plan–specific agreement states only that if the homes are built and the subcontractor performs the work, these are the agreed-upon terms, specifications, and prices.

Material Takeoffs

The purchasing department performs the initial material quantity takeoffs for each plan. The builder and purchasing department collaborate continuously to improve the accuracy of the company's material takeoffs for all their floor plans.

Cost Code	Vendor	Option	Item Number	Qty/ Mult	Item Description/ Contract Number
1590 Windows	2315 84 LUMBER		00V16075	2	2870 FX VINYL LOW E
1590 Windows	2315 84 LUMBER		00V16093	3	3040 FX VINYL LOW E
1590 Windows	2315 84 LUMBER		00V16113	5	3070 FX VINYL LOW E
1590 Windows	2315 84 LUMBER		00V16165	1	4020 FX VINYL LOW E
1590 Windows	2315 84 LUMBER		00V16294	1	3060 SH VINYL LOW E
1590 Windows	2315 84 LUMBER		00V16356	1	6040 FX DL FEB VINYL LOW E
1590 Windows	2315 84 LUMBER		00V16363	1	3060 SH TEMP VINYL LOW E
1590 Windows	2315 84 LUMBER		00V16389	1	4060 SH DL VINYL LOW E
1590 Windows	2315 84 LUMBER		00V16436	2	3070 SH VINYL LOW E
1590 Windows	2315 84 LUMBER		00V16537	2	3070 DL HEB VINYL LOW E
1590 Windows	2315 84 LUMBER		00V16653	1	3070 FX TEMP VINYL LOW E
1590 Windows	2315 84 LUMBER		00V17730	2	2670 FX HEB TEMP RAINGL VINYL LOW E
1590 Windows	2315 84 LUMBER		00V17973	2	1074 FX DL TEMP VINYL LOW E
1590 Windows	2315 84 LUMBER		00004200	1	Builder to Confirm Frame Color
1590 Windows	2315 84 LUMBER	003M-3M	00016000	1	VENDOR TO CONFIRM DL PATTERN PER EACH ELEVATION
1590 Windows	2315 84 LUMBER		00017995	13	Caulk (Window) Tube Sikaflex 219 LM

FIGURE 10.1 Material takeoffs are precise material counts for components of the home according to the plans. Takeoffs are quantity counts not unit costs. With the takeoff made, the unit price may be applied to those quantities to arrive at the total cost of a material item. Seen in this figure is a window takeoff, complete with each item number, item description, and quantity count for each of the windows identified from the plans.

It is the duty of the builder to report takeoff problems such as missing or surplus materials from their jobs to the purchasing department. The purchasing department then adjusts the floor plan's master budget to order less material the next time the floor plan is built.

When there is a material shortage on a job, the builder issues an extra purchase order . If the builder and purchasing group both agree that the extra material quantity was missing from the original takeoff and will be needed every time this floor plan is built in the future, the purchasing department will revise the floor plan's master budget to increase the quantity ordered of that material on future jobs. This process of revising and tightening the material takeoffs based on the first few instances a floor plan is built is part of the cost control and refined efficiency of the production method of building homes.

THE PURCHASING DEPARTMENT'S BROADER RESPONSIBILITIES

The purchasing department's responsibilities include a daunting array of major tasks that reach far beyond the basic bidding and awarding of work. The purchasing group is an integral player in many of the company's most critical functions.

Purchasing department responsibilities are broken down in the following paragraphs.

Estimating Total Cost

Estimating total costs for all active plans is a basic responsibility of the purchasing group. The group is expected to continuously maintain current and accurate cost estimates for all the company's offered floor plans in all its active communities. Accurately estimating total construction costs is essential to the survival of any homebuilding company. Materials and labor conditions are always changing, and a correct estimation of current total construction cost is a large responsibility.

					Total Material			59,897.18	*
3522 Area Dr Accr	9928 ALLOW-W/S	003M-3M	MISLAB				.0000	150.00	N
409 Closing Gift	9928 ALLOW-W/S		OTHERS				.0000	30.00	N
					Total Other			240.00	*
4110 Prod Ovh	9228 ALLOW-W/S	003M-3M	OVERHD	10 Sales Price		551500	0.0230	16,005.10	N
					Total Production Overhead			16,005.10	*
4140 Warranty	9228 ALLOW-W/S	003M-3M	WRRNTY	10 Sales Price		551900	0.0035	1,931.65	N
					Total Warranty Expense			1,931.65	*
							Subtotal	418,676.34	*
							Discount	340.12	*
							Tax	4,131.68	*
							Total	422,467.30	*

FIGURE 10.2 The purchasing department must maintain an accurate total cost estimate for every active plan in the builder's plan lineup. Total cost, as seen in this figure, represents the total lot cost, the cost of subcontract labor and materials, building permits and impact fees, and several other preliminary costs.

Forecasting Future Costs

The purchasing group collaborates with the land/lot acquisition team to accurately forecast construction costs for use in writing proposed new project pro formas. The purchasing group provides the construction costs data that will become important factors in each proposed project's financial projections. Ultimately, the estimated construction costs limit and define the amount the company can pay for new raw land or finished lots. Cost accuracy is critical to securing and succeeding in new communities.

Performing Value Engineering

The purchasing team is a design partner with the architectural group. Costs can drive design, and collaborating with the architectural group during the design development phase of new floor plans helps guide that group as to how a new floor plan's design is—or is not—being designed for a specific retail price target. There are various estimating processes used at different stages of floor plan development to estimate construction costs and then adjust the plan's design early in the design process to guide the home design toward achieving the targeted price point.

System Maintenance

Community and plan setup and ongoing management inside the purchasing system is one of the purchasing department's most important responsibilities. Today's purchasing management systems are rich in features and flexibility to accommodate all the various ways builders prefer to estimate and budget for their jobs. Accurately and quickly estimating jobs and then turning those estimates into POs and subcontracts relies on an extremely powerful purchasing system.

The purchasing professionals who manage big purchasing departments must be—or will quickly become—experts in using and managing the company's job cost budgeting system. Having a great job cost system is key to departmental efficiency and job cost accuracy. Many great purchasing systems are available to homebuilders.

Understanding and being expert at your company's system is critical to running a successful purchasing department. There is more to come on this topic, but the importance of the purchasing system itself cannot be overstated.

Construction Cost Analysis

Cost analysis is an extension of the never-ending process of improving the affordability of your company's homes. How can you give your customers more home value for the same or less construction cost? To answer that question, the purchasing specialist must become a cost analyst.

It is important to continuously analyze your company's construction methods and materials relative to those of your competitors. From this type of competition study, you can make new decisions that affect your home's home design and specifications so to remain competitive. The cost analysis performed during the plan development stage continues after the plan's introduction to ensure that the targeted retail pricing was achieved at the plan's first introduction and maintained over the life of the floor plan.

It is important that the purchasing department continuously provide value engineering feedback to senior management so that the company's desired value proposition and retail pricing strategy can be achieved.

Cost Containment

Construction costs containment and reduction is the joint responsibility of the construction and purchasing departments, with purchasing taking the lead on most initiatives. The process of periodic and new plan bidding gives builders ample opportunity to collaborate with trade partners to revise processes and components, reduce costs, and make the home project as cost-effective as possible for every participant. Construction processes and materials specified are continually evaluated and analyzed as to how the houses can be made less expensive to build or how a specific plan can be brought back in line with the cost of construction of other comparable floor plans.

This continuous cycle of analysis of current costs is a challenging assignment and requires savvy, experienced purchasing experts to properly evaluate, negotiate, and make important decisions that affect each plan's cost and the homes' general quality.

Turnkey contracting versus piecemeal contracting is a good example of where the purchasing group must perform a sophisticated cost-benefit analysis. Performing work piecemeal with smaller, labor-only subcontractors can save money over using larger turnkey subcontractors. Do the higher costs of performing the work turnkey—considering every direct cost difference and production efficiency—result in lowering or increasing total costs? Determining what makes best sense to achieve your company's business plan is a complex decision shared by purchasing and construction.

Component and Installation System Research and Development

Research and development of proposed new component products and/or installation systems is a principal responsibility of the purchasing department. Students who want to see, test, and evaluate the newest building methods and products will love being part of the purchasing department's ongoing construction quest to build better homes.

New Is Risky!

Risk mitigation and avoidance is the other reason for investigating the latest and greatest new products. The purchasing professional must carefully evaluate any new component product or installation method before initiating its use into hundreds or thousands of new homes. Mitigating and avoiding risk in the company's overall construction specifications is the responsibility of the purchasing department. There is a long list of sad cases in which builders spent untold dollars removing and replacing failed component products on the homes that they built several years ago. Being an innovative company is great; being an early adopter is risky.

New building products are constantly introduced, and all homebuilding companies are continuously seeking new and better home components. This ongoing pursuit takes time and effort. Many new products enter the housing market with a burst of excitement, only to fade away in a puff of product liability lawsuits. Does your company want to be an early adopter of new components and technology? There are many good reasons to adopt new, "better" component products, but there also can be pitfalls to doing so. You must be informed and open to new products, but you also must measure your company's appetite for risk and choose your specifications wisely.

Subcontractor Performance

Problem-solving is an important skill for anyone thinking of a career in the purchasing department. The term *construction problem* is somewhat redundant, and the quest for the solution to all those problems is led by the purchasing group.

Periodic Trade and Vendor Performance Assessments

Monitoring vendor and trade performance is an important duty inside the purchasing department. Qualifying prospective and requalifying existing trade and vendor partners along specific criteria to ensure quality, safety, and financial stability are additional demanding duties of the purchasing group.

Formal performance evaluation of the company's trade and vendor partners are designed, implemented, and evaluated and reported on by the purchasing group to the operations management team. When problem partners are exposed, the department and project managers either meet with and provide feedback to those partners to develop them into better partners or replace them with new vendors and trades who appear to better match the company's goals.

Supply Chain Management

Supply chain management ensures that the company's trade and vendor partners have the current and future capacity to meet the company's annual business plan. As the size and complexity of American's builders have grown, the supply chain management role has become essential to ensure that all the company's construction partners can keep up with the company's business plans.

Managing the supply chain involves negotiating longer-term commitments on supply and prices for every home component in quantities that match the number of homes anticipated in the homebuilder's annual business plan.

The supply chain management goal is to anticipate, coordinate, and ensure a ready material and labor capacity with predictable prices for every construction partner extending over the upcoming year or two.

When builders commit to use specific component products, manufacturers offer a per-house or per-component rebate and/or model home discounts on their company's products. In large homebuilding companies, these rebates and model discounts can translate to millions of dollars being returned to the builder.

National accounts rebate management is part of supply chain management and involves measuring and documenting the use of all rebate-eligible products and self-auditing to ensure the rebates are received as agreed upon. The same principles apply to managing the model home rebates and free products offered by those component manufacturers who want their products specified in the builders' homes.

Builders with large land positions and growth plans need to know that the materials and workforce needed will be there to build the homes and make their business plan come true, and that requires managing the entirety of their supply chain.

Creating the Company's Base Home

Writing the company's standard construction specifications and ensuring that they are part of all the company's subcontracts and POs are the purchasing department's responsibilities. Standards and specifications writing is an interesting and important duty that plays a key role in defining the company's home products and even the company's overall market identity.

FEATURE LEVEL		Level 1	Level 2	Level 3
PRODUCT SERIES		BRISTOL SERIES	COSMOPOLITAN SERIES	PROVINCIAL SERIES
INTERIOR FINISH				
	Paint	Monarch	Monarch	Monarch
	Walls	Mopako Flat (7-Colors)	Mopako Flat (7-Colors)	Mopako Flat (7-Colors)
	Ceiling	Mopako Flat (2-Colors)	Mopako Flat (2-Colors)	Mopako Flat (2-Colors)
	Enamel	Semi-Gloss (3-Colors)	Semi-Gloss (3-Colors)	Semi-Gloss (3-Colors)
	Texture	Walls-Spray Texture Ceilings-Stomp & Drag	Walls-Spray Texture Ceilings-Stomp & Drag	Walls-Spray Texture Ceilings-Stomp & Drag
FLOORING				
	Kitchen/Breakfast	Tile "STD"	Tile "A" or Wood "A"	Tile "A" or Wood "A"
	Utility	Tile "STD"	Tile "A" or Wood "A"	Tile "A" or Wood "A"
	Sunroom/Morning room	Tile "STD"	Tile "A" or Wood "A"	Tile "A" or Wood "A"
	Entry/ (Powder, if adjacent to Entry)	Tile "STD" Tile "STD"	Tile "A" or Wood "A"	Tile "C" or Wood "C" (No Slate)
	Powder	Tile "STD"	Tile "A"	Tile "A"
	Family Room	Shaw Carpet - Peachtree Manor	Shaw Carpet - Magnolia Walk	Shaw Carpet - Magnolia Walk
	Closets @ Entry & Hall	Shaw Carpet - Peachtree Manor	Shaw Carpet - Magnolia Walk	Shaw Carpet - Magnolia Walk
	Master Bath	Tile "STD"	Tile "A"	Tile "A"
	Secondary Baths	Tile "STD"	Tile "A"	Tile "A"
	Remaining Areas:	Shaw Carpet - Peachtree Manor over 1/2" Pad	Shaw Carpet - Magnolia Walk over 1/2" Pad	Shaw Carpet - Magnolia Walk over 1/2" Pad
CABINETS	Manufacturer Kitchen/dining (per plan) Std Colors	Quality Oak / Royal, Harvest, Chestnut, Frost Paint Grade	Kent Moore Ash / Paint Grade Autumn Harvest, Golden Nutmeg Paint Grade	Kent Moore Ash / Paint Grade Autumn Harvest, Golden Nutmeg Paint Grade
	Smart Features	Yes	Yes	Yes
	Door Style Wall Cabinets, Uppers	Flat Panel - Liberty 42"	Eyebrow - Sentinel 42"	Eyebrow - Sentinel 42"
	Utility	Per Plan	Per Plan	Per Plan
	Bath Cabinets	Painted Flat Panel Liberty	Painted Raised Panel White	Painted Raised Panel White
COUNTERTOPS	Kitchen	3/4" Granite w/ Polished Edge	3/4" Granite w/ Polished Edge	1 1/2" Granite w/ Waterfall Edge
FIREPLACE				
	Box	Per Plan	Per Plan	Per Plan*
	Surround	Tile	Tile	Tile,Marble
INSULATION	Walls	Polycel soleplates, window, doors R-13 batts w/R3 sheathing	Polycel soleplates, window, doors R-13 batts w/R3 sheathing	Polycel soleplates, window, doors R-13 batts w/R3 sheathing
	Sloped, Knee walls Ceilings	R-22 R-30	R-22 R-30	R-22 R-30
INTERIOR TRIM				
	Interior Doors	Camden - MDF Jambs	Camden - MDF Jambs	Camden - MDF Jambs
	Single Door @ Study,Entry	Camden - MDF Jambs	Camden - MDF Jambs	Camden - MDF Jambs
	Dbl Doors @Study,Entry	Camden w/ T-Astragal (MDF Jambs)	Camden w/ T-Astragal (MDF Jambs)	French w/ T-Astragal (MDF Jambs)
	1 Piece Crown, 4 1/4" MDF	Dining,Entry,Living,Master (Master includes Sitting Room)	Master (Includes Sitting Room)	Master (Includes Sitting Room)
	2 Piece Crown, 4 1/4" MDF	None No Crown - rooms w/Rnded S/R Corners	Dining,Entry,Living Consisting of base & crown No Crown - rooms w/Rnded S/R Corners	Dining,Entry,Living Consisting of base & crown No Crown - rooms w/Rnded S/R Corners

FIGURE 10.3 The purchasing department designs several different feature levels, which allows the builder to offer each active floor plan at various feature levels and price points. The senior team ultimately decides at which level each community should be set. The specification-level sheets are detailed and cover almost every component of the home.

Homes are a fashion product, and older material specifications look dated very quickly. Keeping the material specs fresh and current is essential to staying competitive as homebuyers' tastes are always changing. Tile, light fixtures, flooring, cabinets, and countertops are all hot-button, stylistic components. If these components do not look as good or better than what your customers are seeing on HGTV or at your competitor's model home, then your homes will sell more slowly and for less. That is just the way it is.

Specification writing also includes designing the company's energy package. Today, the most common measure of a home's energy efficiency is the Home Energy Rating System (HERS) score. Achieving the lowest energy consumption score is an artful exercise that involves combining various building methods and component selections to achieve the best overall energy score at the lowest possible construction cost. The combined scores of all your energy-saving elements result in a HERS score. The HERS score can be driven very low (low is good, and high is bad), but at what cost? That is the question that purchasing must answer.

The enormous departmental responsibility of writing the standard specifications is a process, not a project. The company's specifications are never complete because the department is constantly solving problems and adapting to the changing components and methods. Normally, large specification changes are made to coincide with the construction of new model homes. As older model homes are retired and sold, new models are built with the company's newest specifications.

Managing Buyer Selection and Options

Managing buyer selection and options is the purchasing group's responsibility. This role of creating the available finish selections and plan options is a natural extension of writing the company's standard specifications.

The design center has become an important profit center for all production homebuilders. Having a broad offering of selections and options is a must to stay competitive as buyers want extensive choices from today's most current and stylish components, including flooring, cabinets, countertops, hardware, and lighting and plumbing fixtures.

The daily management of the design center operations is probably not the responsibility of the purchasing group; however, managing the selections offered and pricing for every available selection in the design center is.

Plan options such as media rooms, extra bedrooms, and additional bathrooms are buyer options that alter a house's base floor plan. Offering standard plan options is common today and can be extremely profitable. The ability to add the extra cost of a plan option easily and quickly to an individual job's construction budget is an important characteristic of the purchasing system.

Tracking option costs and the resulting option profit margins is also a crucial capability of the purchasing department and its purchasing system. Accurate cost estimates allow for proper pricing of options. It is difficult to distinguish standard job costs from option costs, but this must be done to accurately measure the success of the company's offered plan options.

Scopes of Work

Writing subcontractor scopes of work is an extension of writing your company's standard construction specifications.

Scopes of work:

1. Define the subcontractor's methods of operation (e.g., safety, trash disposal) while on site
2. Define the quality of the subcontractor's work and the allowed tolerances permitted
3. Specifies the approved component products to be supplied by a contractor

Production builders have a written scope of work for every subcontractor category. The written scopes are provided to new subcontractors wanting to bid work and become exhibits in all subcontract agreements for that work category.

When bidding new subcontractors, those bidders are provided with the company's written scope of work for their craft category. This way, you know that a new subcontractor has the information needed to bid your work correctly, with comparable material and labor requirements as those of your current subcontractor. Giving prospective subcontractors your scope of work does not guarantee that they will bid the work to that scope, but it gives them what they need to accurately complete their bid. The purchasing team then reviews and confirms the details and proposed scope of work contained on every bid.

Think of it this way: **The sum of all your scopes of work should add up to the total sum of work required to build a house from beginning to end in accordance to the company's construction specifications.**

Scopes of work are not static documents. They are always evolving, and maintaining current scopes is a time-consuming and ongoing process. Keep in mind that they must always be current!

Bidding and Estimating New Plans

Once plans and specifications are chosen and finalized, then the material takeoffs and subcontractor scopes of work can all be written to match, and the quantities and budgets are compiled into the total cost of construction budget.

Material takeoffs are material quantity counts. The takeoffs are used to create POs, which are utilized to place the material orders to the suppliers. Currently, most material takeoffs are created through manually counting items on the plans by purchasing department team members. Builders also may rely on their vendor trade partners to perform initial material takeoffs. Lumber supply companies will perform a complete lumber material takeoff from your plans for free if you are a regular customer and always buy their lumber for your projects. The quantities provided by vendors or subcontractors should always be reconfirmed in house by the purchasing department's own takeoff. However, using partners to do your takeoffs is a viable alternative if your company has a small purchasing team.

It is worth mentioning that there are independent estimating companies that provide professional estimating services performing material quantity takeoffs for builders. These are good and valuable services that can significantly help a small or overloaded purchasing department. These services are great, but someone at your company still needs to confirm all material quantities and then convert them into POs or into the subcontract agreements.

Some trade partners called *turnkey contractors* bid one lump-sum price to complete their work tasks. The plans and feature specifications need to be sent to turnkey trade partners prior to their bid in ensure that they bid the project per specification. Once the bids are received and carefully reviewed, and if the bids as proposed are accepted, their lump-sum dollar bid amounts are entered into the purchasing system and become the construction budgets for those tasks.

Periodic Rebidding

Rebidding existing costs is a routine duty performed by the purchasing group. Periodic bidding is performed for products whose prices are static and not frequently changing. Costs that have not changed in a while still must be periodically bid on a predetermined periodic schedule, such as quarterly but not greater than annually.

Mechanical, electrical, and plumbing (MEP) or HVAC (heating, ventilation, and air conditioning) subcontractors are usually rebid annually. These subcontractor relationships are important because builders have large and ongoing warranty relationships with those existing subcontractors on all the homes built in the past two years. For this reason, it is critical to manage these annual rebids of your MEP subcontractors with the upmost professionalism.

Purchasing department bidding procedures are a routine part of the accounting audits performed annually by all big companies. Records of all periodic bids and documentation of the decision-making process are always retained for audit purposes. Do not be surprised when auditors ask you to show exactly how all procurement decisions were made.

Opening New Communities

Bidding for upcoming new communities is another opportunity to reach out to all known and qualified bidders to obtain another fresh look at all your construction costs. Bidding for new communities is also an opportunity perform some of the already required periodic bidding when the timing is right. New-community bidding is a reasonable method of documenting procurement decisions in place in other communities, assuming that the current subcontractor continues to be the right decision in the new community.

The process of awarding vendor and subcontractor assignments when entering a new community is a collaborative effort between the project manager, the community's builder, and the purchasing group. These new-community bids also give the company a fresh perspective on its costs that it currently pays within existing communities.

Subcontract Administration

A catch-all phrase, *contract administration*, speaks to the duty and responsibility of subcontractor problem-solving, dispute resolution, and refinement of the scopes of work to match all parties' understanding and intention of the subcontract agreement.

Once jobs are started, the work is occurring, and invoices are submitted, the gaps in scopes of work or pricing differences—if there are any—will be quickly revealed. When there is confusion regarding the bid price or required scope of work, the purchasing team gets involved to ensure the resolution is broadly applied to all other affected jobs systemwide. If the written scope of work is unclear, the purchasing group rewrites it to avoid misunderstandings on future jobs.

The goal of contract administration is to ensure that the scope of work contracted is the work delivered, the subcontractors are happy with the work they did and the amounts they were paid, and any problem discovered is avoided going forward on future houses.

Variance Administration

Takeoff variance administration is a process, not an event. The purchasing group must design the method and criteria for ordering extra materials to job sites. Every extra PO is a budget variance that must be reviewed and approved by the purchasing department and/or project manager, even if the review occurs after the extra materials' order and delivery.

The cost-overrun approval process utilizes a coding system for extra materials based on the reason provided for the extra costs. Correctly coding cost extras initiates the appropriate follow-up action by the project manager or purchasing team.

Suppose, for example, that there was extra framing lumber ordered to a job. The extra PO could be coded to "material misuse" or "takeoff error." If it is coded to the former, there is nothing more to be studied and nothing more to be done at this point. However, if it is coded to the latter, then the purchasing department must step in and investigate whether a permanent takeoff change should be made to the plan master takeoff.

Extras are a big deal in homebuilding companies, and the system for ordering and approving extras is often precise and extensive.

TYPICAL COST VARIANCE CODES

- Approved price increase
- Takeoff error
- Damaged on site—unknown, no back charge
- Damaged on site—with back charge
- Material specification change
- Sales giveaway
- Theft or vandalism

True and honest coding on variances is important for resolving issues on future jobs. The trick is creating a company culture in which the builder's employees are happy to use the variance coding properly. There can be friction between builders in the field and the purchasing team when it comes to coding variances.

Often builder bonus programs reward builders for not having certain types of cost variances. When this is the case, coding to punitive variance codes is probably low, and coding to nonpunitive codes goes up. This results in incorrect coding of extras. Purchasing professionals managing the variance system must keep in mind that proper coding is the goal and that any punitive results from incorrect coding will result in bad data and not accomplish the greater goal of more accurate takeoffs.

The purchasing department tracks extras by variance code type and floor plan and then puts in place automatic steps to be taken when certain variances codes occur and need attention to avoid future problems. The most important variance code needing attention is "takeoff error." If left unresolved, the needed material will be ordered again, with extra POs, and never properly budgeted in future jobs. Not only does this delay future jobs, for which there must be missing pieces ordered, but a builder will never get paid for those materials in their retail home sales prices because those costs are missing from the floor plan's master budget.

Additionally—and ideally—surplus material delivered to a job site should result in the builder submitting a takeoff revision request to lower the quantity of the surplus material for future builds of the same plan.

The contingency budget usually is set as a specific percent of the total construction budget. The amount allocated to this budget varies by the type of home being built. A unique, one-time, custom-built home has a greater likelihood of and tolerance for extras than does a repetitively built production home. A custom home budget can have many allowances, rather than firm budgets, and usually has a generous contingency budget.

Production builders often will set up a contingency budget of about 2 percent of total direct construction costs—not of the sales price, but the construction costs.

Well-run production purchasing departments and construction teams can see results of about a 1 percent variance from the original construction budget as an annual average across a large number of jobs built. Some jobs may have no extras, whereas others have a high number of extras. It is the overall average percentage variance result that indicates the company's budgeting performance.

Start Packages

Producing the starts packages is often the responsibility of the purchasing department. In some organizations, the starts function falls into the architectural group, or there is a standalone starts department. However, wherever the starts process responsibility resides, the purchasing department has a large responsibility relating to the production of the individual job's unique budgets contained in its start package.

Each house or job must have its construction budget individually adjusted to match the buyer's selections and options and accurately reflect all the costs associated with the unique lot conditions for that job. A buyer's selections and options are usually pre-estimated, with preset budgets for easy input by the starts team when those options are selected. However, the lot unique budget adjustments cannot be estimated from the office; therefore, all unique changes to each job's budgets that are affected by the lot conditions are submitted by the builder assigned to that job and input into the job's budget by the purchasing team.

The objective of adjusting all affected cost categories is to provide the builder accurate line-item budgets for every needed payment on the job, resulting in a variance-free job. Producing variance-free starts is a highly aspirational goal—but it is still the objective.

PURCHASING IS A HIGH-PROFILE DEPARTMENT

Homebuilding companies live and die by their gross profit margins. There is no business metric more closely monitored in the homebuilding industry than the gross margin results, to which the purchasing department is closely tied. Fair or not, the purchasing group often receives the credit—or the blame—for the company's earned gross margins.

If the current market prevents a homebuilder from raising its sales prices enough to improve margins to the business plan's target, then the burden of achieving the target gross margin falls on the purchasing group to reduce construction costs until the target gross margin is achieved.

The thinking is that if competitive homebuilders are happy with their margins and selling their homes at those low sales prices, then why can't we receive good margins by selling our homes at those same low sales prices? The logic follows that this low gross margin problem is a high-costs problem and therefore must be a purchasing problem. This delicate balance between being a hero and being the problem is the life of the purchasing manager.

Good gross margins and accurate budgets are a requisite to being a successful homebuilding company. For this reason, the position of purchasing manager has risen to a senior management level in most homebuilding companies.

PURCHASING DEPARTMENT ORGANIZATION

The purchasing department is led by the purchasing manager or director of purchasing, and in some companies, the purchasing role has been elevated to the vice president of purchasing. These titles vary by company based on the volume of homes built and the company's internal title nomenclature. The purchasing manager oversees a team of purchasing specialists who collectively perform all the responsibilities as described in this chapter.

Purchasing specialists may be assigned responsibilities based on a couple of arrangements.

First, each specialist may be responsible for a specific group of material and labor cost codes, for which the specialist must ensure current and correct prices and accurate takeoffs and specifications across all plans and all projects for those assigned cost codes.

For example, a purchasing specialist may be assigned responsibility over windows, doors, and interior trim. In this example, the employee manages all matters related to those items. If there is an extra window ordered and that extra is coded to "takeoff error," that window specialist researches the window extra and determines whether a permanent takeoff change to the plan's window takeoff is needed. If one is needed, the specialist makes that takeoff change in the system and thus avoids the problem in the future. Additionally, if the window vendor comes in to talk about a price increase and/or other changes related to the windows, that vendor will meet with that window-purchasing specialist. If a new window company wants to bid your work, the specialist coordinates the new window company's bid, evaluating the window company itself and its window products, services, and reputation before considering bringing them on to the approved vendor list. This specialist must serve as the expert in all things about windows, doors, and trim and run the procurement for the assigned components.

The second organizational design entails one in which purchasing specialists are responsible for managing costs and takeoffs by assigned trades and vendors. If, for example, vendor ABC Inc. sells five products or services, then the specialist assigned to ABC, Inc., covers those products—thus avoiding a situation in which a vendor would have to deal with two or more specialists on all the items or services they offer.

Under both of these arrangements, the sum of all the specialist assignments must cover every subcontract and material purchase that is required to build the houses.

One last and less common purchasing department structure is one in which a specialist is assigned total control over all cost codes for a specific project or group of communities. In order to require the use of this system, a community must be extremely large and active, offering a variety of home types. One such community is Kissing Tree in San Marcos Texas, where there is an on-site purchasing department dedicated exclusively to the community.

Some purchasing employees prefer to manage total project costs rather than category costs. There are advantages and disadvantages to each of these department designs.

This professor prefers the cost code category specialist design, in which the specialist is focused and expert on specific components and subcontract categories. It is hard to be an expert in every single cost code! The cost code specialist can ensure that the best components and subcontractors are used across all the company's communities for their assigned components.

The Purchasing Manager

Purchasing managers are among the most difficult people to hire and retain in the home-building industry. The reason for this is complex. The purchasing department's challenge is to attract the best construction experts who also enjoy working on computers in an office environment. The challenge of being a great purchasing manager is trying to be an analytically inclined construction expert who also happens to have strong people management skills. That unicorn of an employee is in high demand and exceptionally difficult to find. If CSM students fit that description, they will have a great career in purchasing department management.

Open purchasing manager positions found on Indeed.com or Monster.com provided some interesting insight into the duties and responsibilities for the open purchasing manager positions found on any one of the common employment sites.

PURCHASING MANAGER DUTIES AND RESPONSIBILITIES

- Manage the entire bid process.
- Negotiate and finalize the prices and details of all contacts.
- Manage all subcontracts and purchase order documents.
- Audit and analyze the performance of existing trades and suppliers.
- Maintain the company's standard specifications.
- Provide accurate direct job cost estimates to the land acquisitions group.
- Maintain current scopes of work for all subcontractors.
- Maintain accurate total house construction costs data.
- Attend frame walks and provide pricing and product expertise to the design process.
- Perform research and resolve trade partner disputes.
- Review and approve construction cost variances.
- Provide in-depth cost analysis and value engineering for all plans and construction methods.
- Continually source new and improved component products.
- Continually source new installation methods.
- Continually source new subcontractors.
- Maximize rebates by negotiating and securing the best available rebates and build future trade relations opportunities.
- Actively participate in the creation and implementation of the annual business plan.

SUMMARY

The purchasing department is the research and development arm of the company. A well-run purchasing department budgets and delivers at-or-below-market construction costs while seeking everything new for the future of tomorrow's home. The methods by which homes are built and the components that can change a home's integrity and style are all discovered by the purchasing department.

The aspect that this professor enjoyed the most about working in purchasing was the opportunity to be part of the development of the company's future home products. The purchasing department is a great platform for finding the future of the home and the future for your company and your career. The purchasing department is about finding the future for the American home.

REVIEW QUESTIONS

1. What are some reasons to spend at least a few years working in the purchasing department?
2. How are new subcontractors brought on board? How can builders get their new subcontractors administratively approved and ready to work on their job?
3. Describe how the subcontract agreement system works. Do subcontractor agreements lock the builder or the subcontractor into a specific number of homes or work-specific addresses?
4. What is meant by the term *material takeoffs*? How are takeoffs obtained and managed?
5. In what ways does the purchasing group collaborate with the other departments inside the company? How does the purchasing manager interact with the land/lot acquisition group and the plan development team?
6. What does it mean when it is said that the purchasing department is the research and development arm of the homebuilding company?
7. Why are the standard construction specifications and scopes of work so important to the homebuilder? How do they define the builder's product?
8. Define *periodic bidding*. How might this required bidding be accomplished, and what purpose does it serve? What types of products or services are bid periodically?
9. Describe some of the common cost-overrun variance codes used by homebuilders.
10. What are contingency budgets, and what size should be anticipated/accepted for the various types of houses you might build?
11. What makes the purchasing department such a high-profile department? What is *gross profit margin*?
12. Describe the various ways in which workflow inside the purchasing department may be designed or assigned, and why.

Figure Credits

CHAPTER 11

IT SOLUTIONS FOR HOMEBUILDERS

INTRODUCTION

Homebuilders are working to find new applications for automation throughout their processes. Huge strides in process automation are being made across all departments. The industry and its users have been slow to adopt and adapt, but that is changing quickly inside small and large homebuilders alike. What once were a collection of disconnected processes and systems are becoming linked in real-time by better software developed specifically for homebuilding. Construction Science and Management (CSM) students who excel at construction and computer science may find an interesting career path previously not envisioned. In this chapter, we look at an overview of the role of process automation and information technology (IT) solutions for homebuilders.

HOMEBUILDING'S AUTOMATION EVOLUTION

Homebuilders have been slow relative to other industries over the past 50 years in applying new technology to their operational construction and sales processes.

Early automation was focused primarily on following the money; accounting systems were first followed by and connected to purchasing systems. Early accounting systems took the lead, and so, today, the accounting system remains the backbone of many companies' entire IT system.

Automated purchasing systems were next because they fed the accounting system with the budgets needed to complete the money loop. These accounting systems teamed with the purchasing system's revolutionized job cost budgeting, accounting, and control processes.

Accounting and purchasing systems working together brought large changes and improvement to homebuilding operations and quickly produced better financial results. For years, builders seemed satisfied with these two systems working together for faster, easier, and more accurate budgeting and accounting and the job cost control benefits that the two systems together delivered.

Job Cost Management, 1980s Style

It is hard to imagine, but when this professor entered the homebuilding industry in the early 1980s, our material takeoffs were in three-ring binders, with one copy kept by the material supplier, one copy kept by the builder in the field, and one copy kept in the main office. With these three binders, we attempted to manage and control our material orders and payments.

When the builder released a material package by phone for delivery, the vendor went to their binder and shipped the specified material in the quantities indicated in their version of the book. When the vendor delivered the material, the builder was supposed to confirm the delivered quantities of material against what was specified in their field copy binder. Then the vendor invoiced for the delivered materials, and the accounts payable group received the invoice, looked up the plan number in their binder copy, and confirmed the material shipped matched the material type and quantities itemized in the office binder. That should have worked, right?

Well, the notebooks and invoices never matched. It was a nightmare for all parties trying to get materials delivered and invoices paid. If the invoice was even close to our budget, we paid it and kept on moving. I do not remember every seeing a matching invoice and three-ring binder page takeoff!

Extra purchase orders were issued in the field, and so those invoices arrived without warning or budgets waiting in the system. Extras were the wild card! There was no good system to document or predict or control extra orders. Builders had to rely on their contingency budget and hope that was enough to cover our dated and inaccurate takeoffs and obscure extra ordering processes.

And as far as managing our takeoffs with the needed changes between all three of the binders, forget it! There was no way to keep all three binders current and matching all the time. We just did our best, and the strong housing market made up for our paper systems. That was the early 1980s' production building method of budgeting and job cost control.

Automating One Process at a Time

Builders initially automated their various operating processes in construction, warranty, sales, and other departments by selecting independent systems for each of the processes they wished to automate. This one-process-at-a-time approach to automation started with the selection and implementation of their accounting system. Because the automated accounting system was there first, the accounting system became the backbone system to which other automated processes were later connected. This is how it started, and, in some companies, this is how it still is.

Companies quickly adopted and adapted to everything around their automated accounting systems. Homebuilders became committed to their accounting system. When new automation opportunities presented themselves, such as a warranty, sales, or scheduling, those systems were initially looked at like options that could possibly be purchased and bolted onto their accounting system to which they were already fully committed.

Early Job Cost Management Automated

Early job cost accounting systems were generic and aimed at any and all construction companies from any construction sector. Early systems allowed the builder to create a unique job and provided a place for the contractor to input cost coded, line-item budgets and track posted payments against each payment's budget. Variances were displayed as positive or negative with columns for the budget, the payment, and the variance, and that was it.

Even as basic as it was, that simple automation refined and improved financial control for homebuilders of all sizes. These early systems gave homebuilding executives and company owners the fastest and best financial information they ever had. The accounting systems led the awakening of the builders to the benefits of automating financial information.

Job cost accounting systems continued to improve and shine a bright light on the company's job costs, estimating (in)accuracy and job cost control in the field. Line-item visibility on every cost code's budget against actual costs revealed the huge opportunities for improvement that were awaiting. This visibility on the new opportunities ramped up the focus on improved estimating systems and job cost budgeting and control solutions.

Next-Generation Systems: The Plan Master Budget

The next generation of purchasing systems for production homebuilders provided something they desperately needed: the plan master budget.

These new systems allowed builders to build and maintain a master budget for every floor plan they offered. That master budget allowed builders to not just input a budget but also itemize materials; sort takeoffs by community, elevation, and swing; and then generate the current construction cost estimate accurately and instantly for every plan offered and, in every community, where it is offered. Wow!

Further, the new systems allowed a plan's master budgets to be copied into a newly created job, an address-specific ledger. Once the master budget and its contracts and material takeoffs had been copied into the job, that job's budgets could then be customized as needed for customer selections and lot conditions changes without changing the plan master budget. Job-specific changes were able to be made to the job and not affect the plan's master budget.

Another important feature of these new systems was that they provided for easy maintenance of each plan's master budget, material takeoffs, and subcontract budgets. Managing these systems is a learned skill set that today's CSM students will easily acquire. Costs and floor plans are always changing, and easy maintenance is an important characteristic for all purchasing systems. Learn all you can about your company's system.

Each floor plan's plan master is maintained to be accurate at all times, ready to start the next house today. To start a house, the system opens a new job number for a specific address. The plan master's current takeoffs, unit prices, and subcontract amounts are all copied into that newly created job file. Future payments made to the line-item budgets hopefully match, but if not, the system provides a cost variance reporting and approving process that greatly improve the builder's ability to see trends and avid problems in the future.

By refining and improving their accounting and estimating systems over the decades, builders made huge strides in improving their job cost budgets and control of the line-item budgets from the field. CSM students should anticipate working on these systems upon entering the industry. Although these were big steps for builders, there was much more automation and improvement yet to come for homebuilders.

BEYOND ACCOUNTING AND ESTIMATING SYSTEMS

Today, homebuilders are seeking and finding great solutions for all their accounting, purchasing/estimating, architectural, construction, warranty, document storage, and client relations management processes. Integration between all the various financial and operating systems has already changed every homebuilding process across every department.

Bolting Separate Systems Together or Single-Source Solution

CSM graduates entering homebuilding should seek to understand as much as they can about the system used at all the homebuilders where they might work or do an internship. They are likely using a variety of separate systems to meet their different needs. How those systems are built and communicate with each other is different from system to system. This professor is not a systems expert, but this chapter provides an overview of what I have seen and learned over the last many decades.

Ready for More

Homebuilders are very ready for further process automation. CSM students who are interested in construction and IT will find a lot of opportunities to develop their career by blending those two skill sets. Today, many single-source solutions are available to homebuilders. These single-source solution companies integrate many departmental systems, allowing unprecedented access to accounting and operational data to employees and other people inside and outside the company. These new-generation systems even provide outside suppliers and contractors limited visibility on payment history and other project-relevant information. Today, the integration of all these systems is opening up all kinds of new methods of opportunity for improved efficiency and effectiveness inside homebuilders.

ARCHITECTURAL SYSTEMS

Another early automation revolution along with the accounting and purchasing systems was the automation of the construction drawings, from hand-drawn plans to computer-assisted design (CAD) drawings. Of course, this extremely important innovation was not specific to homebuilding; the positive benefits of CAD extend into every construction sector.

This professor worked with all hand-drawn plans until about the mid-1990s. The transition of all plans into CAD drawings was a monumental task with monumental results. As the pace of plan production increased, so, too, did the work pace for other departments. Fast plan production and changes sent work to the purchasing department faster than ever; as you can imagine, speed in one department affected the need for improved automation in all the other departments.

Computer-Assisted Design

The move to CAD drawings changed the architectural world forever. Maintaining accurate plans and making floor plan revisions or creating new plan versions of old plans became almost simple compared to the labor-intensive process of creating and revising hand drawings.

CAD drawing's new change of pace had the most impact on the purchasing department. If architecture can change the plans with a double click, how was the purchasing department supposed to keep takeoffs and budgets current using manual estimating methods? The move to CAD drawings was a revolution that rippled throughout the entire company.

Three-Dimensional (3D) Architecture

CSM students are familiar with building information modeling, or BIM. The next wave in automated architecture came to homebuilding in the form of 3D plans, essentially BIM for homebuilders. Please note that 3D architectural systems have not been universally accepted in homebuilding. Why not?

True 3D plans may offer many advantages and applications for homebuilders. Builders and buyers can walk through a home virtually without even building it and see the house fully furnished and decorated in 3D. This is great and offers much potential for designing new plans and for selling unbuilt homes. These advanced 3D systems also offer material takeoffs for every component in the home. Automatic and accurate material takeoffs, wow. What great and time-saving advancement!

Three-Dimensional Drawn Homes: The Future, or Not?

Why are builders not already using advanced 3D CAD architectural and estimating systems today? The answer lies not in the lack of great benefits that could be realized once the system is fully operational, but rather in the great effort it takes build and maintain this system.

The 3D architectural system must be complete, containing every possible component of a home built into that system by a unique item number to be placed onto a plan by the architect or CAD operator. Basically, this system requires building the house piece by piece in the architectural system. This architectural process, in its current form, may be even more work than building the house in the field.

This professor once spent over $1 million implementing the latest and greatest 3D architectural system for our company only to soon turn it off and abandon the new system and the million dollars completely and return to our CAD two-dimensional (2D) AutoCAD system.

Why did we walk away? It was it taking too long to produce a new set of architectural plans. The time it took to produce plans did not make up for the time saved by having to do manual takeoffs. The new system simply moved work from the purchasing department to the architectural department, and there was no net savings of time.

Additionally, we could not find enough system operators to keep up with our workload. We could train someone, but if that person left, everything we invested in their training was lost. Essentially, we could not find, train, and retain qualified 3D system operators as we could when using the 2D CAD system that everyone knew and could operate on day one on the job. It is probable that one day soon the 3D systems and the operators will mature and become more common and the obvious system of choice, but for now, it is rare you will find a production builder drawing 3D plans.

CONSTRUCTION SCHEDULING SYSTEMS

Homebuilders have been creating various forms of scheduling systems forever. The construction sequence and likely date schedule for a home can easily be predicted at the start of the home project.

That initial schedule is subject to change every week. Old-school builders are always planning out their next few weeks' schedule using their paper system that someone showed them when they first got in the business. That is where it started, and that is where it still is with many builders, even where their company has a great schedule system. They just prefer to do it manually. This professor predicts that today's CSM graduates will embrace these systems and use them to their maximum potential to produce excellent results.

Passive Schedule Systems

Early automated scheduling systems were passive systems. These systems utilized named stages of construction and created weekly progress reports based on manually input stage changes by the builder.

The system then created a weekly status report, which told project managers which houses were moving stages, and which were not. If a house was at a stage 5 the previous week and the builder updated the job stage during the current week to a stage 6, it was called a *making a move*. The idea was that if a house is making moves, or following the assigned schedule, then all was well. However, those homes not making a move for one or two or three weeks triggered an exception. The system reported nonmoving houses to the project manager, telling them which houses were stuck for some reason in a certain stage. Project managers could manage their builders by looking first at their stuck houses and not worrying as much about all the other houses that were making moves each week. In this professor's day, we called the stuck house report the *red-flag report*.

This type of system was passive and did nothing more than report the moves each house made or did not make. The only proactive element was the creation of various reports to get the passive information into the correct hands.

Proactive Scheduling Systems

The newest systems do more than just report the schedule status; they read the schedule and prompt next actions required by the builder based on the home's current status. These systems contain the sequence of tasks and lead times for each task. Based on the status of the home and identified lead time for the next task, the system prompts the builders on every next task needed on all their active homes by producing the builder's daily to-do list. These systems can alert and remind the builder of needed material releases and subcontract notifications, prompting the builder on every next step in the construction process. These systems can eliminate the need for the builder to routinely perform the paper scheduling process.

CSM students should be eager to learn the many features of today's scheduling systems. Today's CSM graduates will take full advantage of these new systems and become even better builders than their predecessors, the accomplished paper-schedule builders. CSM students should seek out training and embrace whatever system is used by their homebuilding employer. Using the scheduling system will help you gain confidence early and become an even more efficient and effective homebuilder.

Two companies offering great homebuilder software are Hyphen Solutions BuildPro and Constellation Homebuilder Systems. Check out their websites for information about homebuilding automation.

WATCH AT https://tinyurl.com/44vs928k

WATCH AT https://tinyurl.com/yxh2caw

WARRANTY SYSTEMS

Warranty systems maintain records and create valuable reports based on completed and pending warranty activity. These systems monitor open warranty cases through completion.

Records retention and open call status are the two most basic requirements of any warranty system.

Today's newest warranty systems also provide homeowner portals through which homeowners can initiate or update warranty requests to the builder and find helpful information about the terms of their warranty, their color selections, and other helpful homeowner information the builder provides to all their customers. The builder can also use these portals to communicate with homeowners about specific maintenance dates on their home, precautions to be taken in bad weather conditions, or any other home-relevant items.

Warranty is a much-overlooked aspect of residential construction. Homebuilders have a builder-provided limited warranty and other legal obligations on their homes for up to 10 years from the date of the home closing. A complete history of each home must be maintained by the builder in the event of future problems.

House-Specific Records

Difficult buyers and/or warranty situations must be managed with complete professionalism. Detailed records must be kept in the system on every letter, email, phone call, and any other third-party communication regarding the house. Warranty representatives are trained to keep detailed notes in the system about what the homeowner says, believes, and expects to happen next from the builder. Future warranty representatives to this house need to know the complete history of the house and homeowner.

Builders even keep records on the homeowner's temperament and write down advice or clues to future representatives about how best to meet this customer's expectations and ultimately make that person happy. Additionally, any specific messages that were imparted to the customer by phone or in person must be documented completely. Email records are always available but are also saved to the warranty system in these cases.

Recurring Warranty Issue Records

Warranty spending is the most expensive spending in homebuilding. When you are spending money on fixing problems in your homes, you are also spending goodwill with your customers and spending your employees' valuable time. And all that money and time spending generates zero profit! Warranty effort and cost are just time and money lost forever.

Production builders repetitively install the same components using the same installation methods hundreds or thousands of times a year. Warranty systems extract data and trends about material and workmanship failures across all your homes. Sophisticated builders take those data and their indicated trends and act to replace faulty components or revise or reject failing installation methods.

There is a fortune of problem-solving information that can be found in the company's warranty spending and home visit patterns. Frequent, small, and cheap-to-fix problems are just as important to avoid trips to homes and inconvenience to homeowners. Builders need data on problems of all sizes. Small, repetitive problems can be rooted out from the data in your warranty system.

The aim of the warranty system is to give the builder warranty trend information to avoid reoccurring, future warranty issues and to track and control open warranty items and thus build and ensure good relationships with your homeowners.

CUSTOMER RELATIONS MANAGEMENT (CRM) SYSTEMS

CRM systems create a company's initial contacts with new prospects and follow them as they become buyers and then homeowners and part of the warranty system. These integrated systems are among the latest of the systems adopted by builders. CRM systems were initially and primarily designed as a sales management system but have evolved into much more than just a prospect file follow-up system.

Today's CRM systems for builders are designed to create a record for every person with whom the company comes in contact. The CRM system follows that individual or family from that first day of the initial contact and follows that client throughout the relationship to becoming a homeowner.

The capabilities and benefits that may evolve from the CRM system are endless. Home shoppers may someday prefer buying the home completely remotely, and the CRM sales and client relations system provides the platform to interact with home buyers throughout their new home experience.

Sales Systems

The CRM was the early sales management system. These early CRM systems prompted the salesperson to initiate recontact with prospect and record all contacts with each prospect. These systems were primarily designed to provide sales managers with activity reports on all their salespeople. The generic systems were great sales management systems for all kinds of industries but did not satisfy all the homebuilder's needs.

Today, builders use industry-specific systems that provide for monitoring a wide variety of sales-related activities in homebuilding. Salespeople can write and sign sales contracts with buyers on the system and track various necessary steps that must be completed by the buyer or salesperson to advance the sale through the builder's process regarding the mortgage loan approval, the status of needed deposits, or any other actions that progress the home through the builder's system.

SUMMARY

Production builders have come along way over the past 30 years, but there is still a lot of automation needed in the industry. The benefits of what has already occurred are clear, and builders continue to seek new ways to improve and evolve. CSM graduates entering the industry will find plenty of room left for them to make their mark on automating the industry.

REVIEW QUESTIONS

1. Which were the first two existing systems to automate their processes? Why do you think those two were first?
2. Describe the concept of the plan master budget. What features and benefits did this next generation purchasing system provide homebuilders?
3. Describe the one-process-at-a-time approach that builders followed in process automation. Why were most of the new systems built around and onto the central accounting system?
4. Why are most builders still using 2D CAD rather than 3D CAD architectural systems?
5. What is the difference between passive and proactive construction scheduling systems?
6. What kinds of records are kept for the builder and homeowner by the warranty system? How does the warranty system monitor open warranty calls? What is a homeowner portal?
7. Why might builders want to have companywide data on the material and workmanship performance issues they encountered during their home's warranty period?
8. What is the CRM system? How is this used in the sales offices and beyond, during the life of the buyer relationship?

THE STARTS PROCESS

CHAPTER 12

INTRODUCTION

Reaching the point where a residential construction project is ready to start construction is an extraordinarily complex process. Significant preparation, decision-making, and architectural and budgeting effort are required to secure a building permit and deliver to the field everything needed to build the house on specification, on budget and on schedule. The starts process is the first step in the construction process that builds a new home.

The Starts Department

The starts process is so important and complex that in large production homebuilders, it is assigned to a starts manager who runs the starts department. Imagine all the work to start one house, times 500 or 1,000 or even 2,000 starts a year in a single city!

Managing so many homes in the starts process can be a crushing responsibility and requires good systems and constant management to keep the starts flowing out of the corporate office and into the field.

The company's annual business plan relies on a specific number of home starts occurring in the scheduled month shown in that business plan. If you do not get the planned starts out, you cannot achieve the planned closings. Right? Therefore, in large companies, the starts process is likely to be its own group, a team dedicated to nothing but pumping out the starts to match the starts detailed in the annual business plan.

In smaller homebuilding companies, the tasks involved in starting their homes are likely distributed around the corporate office and not to a single department. Either way, all the steps are the same in large, medium, or small companies; the starts process must be completed before construction may commence on any home.

Starts Manager

Construction Science and Management (CSM) students may someday be asked to lead a starts group. An opportunity to manage this group would be a large responsibility and provide valuable experience and exposure inside the main office. Leading this group may not be your dream career endgame, but it is an important position in any production homebuilding company and a worthy place to spend a few years.

The starts manager leads the weekly starts meeting, which managers from the architectural and construction teams attend. At these meetings, every planned start in the system is reviewed individually, and action items are assigned to managers as needed to ensure every start progresses through each step required to deliver an accurate and complete start package to the builder in the field.

The starts manager's is charged with achieving the company's prestart cycle time goal. On sold homes, the start cycle is usually measured from date of sale to date of start. This metric is measured from the sale contract acceptance date to the day the complete starts package is delivered to the field. Spec homes' starts cycle is measured from the date the job enters the starts process report until the day that start is delivered as a final starts package to the field. The number of average days to start a sold or spec house is an important and common homebuilding metric.

THE STARTS DEPARTMENT DUTIES

The starts process includes many tasks that are performed inside the main office and other duties that are performed in the field. CSM students will likely begin their career in the field, so please pay special attention to these duties that performed on your behalf by the starts department in the office. You will not have clear visibility early on your career on all these activities as a builder in the field, but you need to understand those office tasks as well.

The builder in the field has their own responsibilities related to completing the starts process. **The field responsibilities include:**

- **Evaluating the actual lot conditions as compared against the preliminary plot plan** and recommending any changes to the plot plan as needed to improve the house fit or avoid obstacles such as manholes or electrical boxes.
- **Evaluating the lot conditions relative to the standard house budgets.** The builder must estimate unique costs associated with this specific house being built on this specific lot. Costs associated with topography, native trees, and other existing lot conditions must be accounted for and budgeted into the starts package correctly. These lot-specific budgets come from the builder.
- **Submitting all recommended changes to the corporate office.** This includes recommended plot plan changes, if any, and construction budget changes to the purchasing department as needed to build the house as planned. Most companies have a formal process for submitting lot-specific costs for budget adjustments prior to the start.
- **Working with the salesperson** to facilitate the buyer completing the options and selections decisions promptly, including answering any buyer questions regarding the home or the lot or performing any requested research or whatever can be done to facilitate the buyer's selection and options.

The corporate office responsibilities include:

- **Job-specific plot plan:** Creating the final version of the plot plan, from early versions that used the plan and lot templets and placing the house and flatwork on the lot as approved by the buyer in the early versions. Buyer approves and signs the final version.
- **Job-specific plan set:** Architectural and engineering plans are revised specific to the buyer's selected plan options or other needed plan changes.
- **Buyer-specific budgets:** These are input based on buyer or design center selections and plan options.
- **Lot-specific costs budgets adjusted:** Lot-specific budgets are written in the field and input into the job budgets by the purchasing team.
- **Job-specific specifications:** These include the buyer selections summary and more.

NOTE: Every field and corporate office task (listed above) must be completed before the permit and homeowners' association (HOA) approval applications (listed below) can even be submitted.

Permit and approval steps include:

- HOA/Community approval
- Developer plan review approval (if any)
- Municipal or county permitting
- Obtaining utilities, taps, meters, etc. (only with permits in hand)

The starts package is comprised of all the HOA approvals and municipal permits, job-specific budgets, architectural- and engineering-stamped and city-approved drawings, various company-specific documents detailing buyer selections, and other job-specific information required by the builder during construction.

BUYER MANAGEMENT DURING THE STARTS PROCESS

Managing the buyer through a successful starts process is key to an on-time construction start and, ultimately, a satisfied client. Production builders are selling their homes from a model home, offering both move-in-ready spec homes as well as vacant lots in which buyers can choose their lot and floor plan. Once the sales contract is written, the race is on to get the home started before construction costs increase and the buyers change their mind.

Job-specific plans need revising, and job-specific budgets take time to estimate and input into each cost code budget line item affected. This time factor puts pressure on the builder team and the buyer to decide on and pay for all selections and upgrades as quickly as possible. A buyer who cannot quickly decide if they want the media room option or not is holding up the entire starts process for their home.

Production builders sell their homes at fixed prices based on their current construction costs. However, the builder's costs are not fixed. In fact, costs are always on the rise. Every team member

Pre-Construction Meeting

Date:__________________

Address:______________________________

1. Plot Plan – Verify correct lot_
   a. Review orientation of home on lot and buyer to initial plot plan
      i. Centered_ Left_ Right_
   b. Type of Drainage on lot
      i. A) Front to back_ B) Right to Left_ C) Left to Right_
      D) Back to Front_ E) Other_
   c. Any utility boxes on lot?
      i. Mark location on plot plan if not already identified_
   d. Any lot requirements by Developer, City or other Government entity?
      Please note here____
   e. Note approximate fence location_

2. Architectural Plans – Verify correct plan_ - Verify correct elevation_
   a. Review all elevations and buyer to initial page with front elevation_
   b. Review any and all structural option selections_
   c. Review any and all mechanical option selections_
   d. Review any and all NSOs (non-standard options)_

3. Selection Sheet –
   a. Review selection sheet for construction related items_
   b. If there are any questions about their selections refer buyer to the NHC.

4. Set build and process expectations
   a. Buyer will be updated at each stage move of the home by the NHC (give copy of stage move sheet)_
   b. ALL communication from and to buyer will be directed through the NHC and not CM_
   c. Buyer to request permission from NHC prior to visiting jobsite_
      i. Do not visit jobsite during normal business hours or when workers are present_
      ii. Do not add anything to the home during the construction process_
   d. There are no changes allowed, at any time_
   e. The home is a construction site (cleanliness expectations)_
   e. Completion date to be given at Stage 7 of the home and not prior_
      i. Stage 7 is Countertops installed, Exterior Masonry Complete and Underground Gas Line to home _

5. Discuss Eliant Move in Survey

Construction Manager:______________________________

Buyer:______________________________

FIGURE 12.1 Production builders meet with each buyer immediately before starting construction to build rapport and trust with the buyer for the construction journey that lies ahead. The main purpose of the meeting is to set the buyer's expectations to match the builder's established process.

in the starts group, the design center, and the sales office must work systematically to present and secure the buyer's option and selection decisions and required deposits from the buyer as quickly as possible so that the process of planning and budget revision may begin.

The company's selections processes and systems at the design center are key to completing this stage as quickly and successfully as possible.

The Plan Review Meeting

Production builders that offer plan options usually conduct a plan review meeting with the buyers. This is a formal meeting during which every page of the plans and every offered plan option is reviewed carefully with and offered to the buyers. These options may be small, such as adding an electrical wall plug, or extensive, such as adding a media room, bathroom, or bedroom.

The goal of the plan review meeting—builder's technology allowing—is to completely finalize the options selected with deposits and accordingly revised the plans during the meeting and obtain the buyer's signature(s) on that revised plan set. This signed plan set becomes the permit plan set and construction plan set, complete with the buyer's signature.

The Preconstruction Meeting

Immediately preceding the start of each sold home, the builder meets with the buyers to review the construction process that their home will follow, as well as the lot and plot plan and confirm all options and selections one last time. The underlying theme of the preconstruction meeting is to properly set the buyers' expectations to match the builder's actual process. When buyers understand the process and the builder follows their own process, the result will be a happy customer. Most production builders follow a set preconstruction meeting agenda that helps ensure that every important point is covered with every buyer. The agenda of the preconstruction meeting is often signed by both the builder and buyer and then kept in the job's saved documents file.

THE JOB-SPECIFIC PLAN SET

The Job-Specific Plot Plan

Placing the home on the lot is one of the earliest and most important decisions to be made during the starts process. Every lot is three-dimensionally unique regarding its topography, shape, and size. The builder in the field must evaluate existing improvements such as manholes or storm inlets, community mailboxes, streetlights, existing native trees, ground or aerial easements, and other community-specific requirements to determine if the home may be placed on the lot in according to the initial plot plan. Next, the more esthetic considerations come into play, such as special views, sun orientation, garage swing preference, and backyard size and orientation. The possible combinations and arrangements of a home on a plot plan are almost endless.

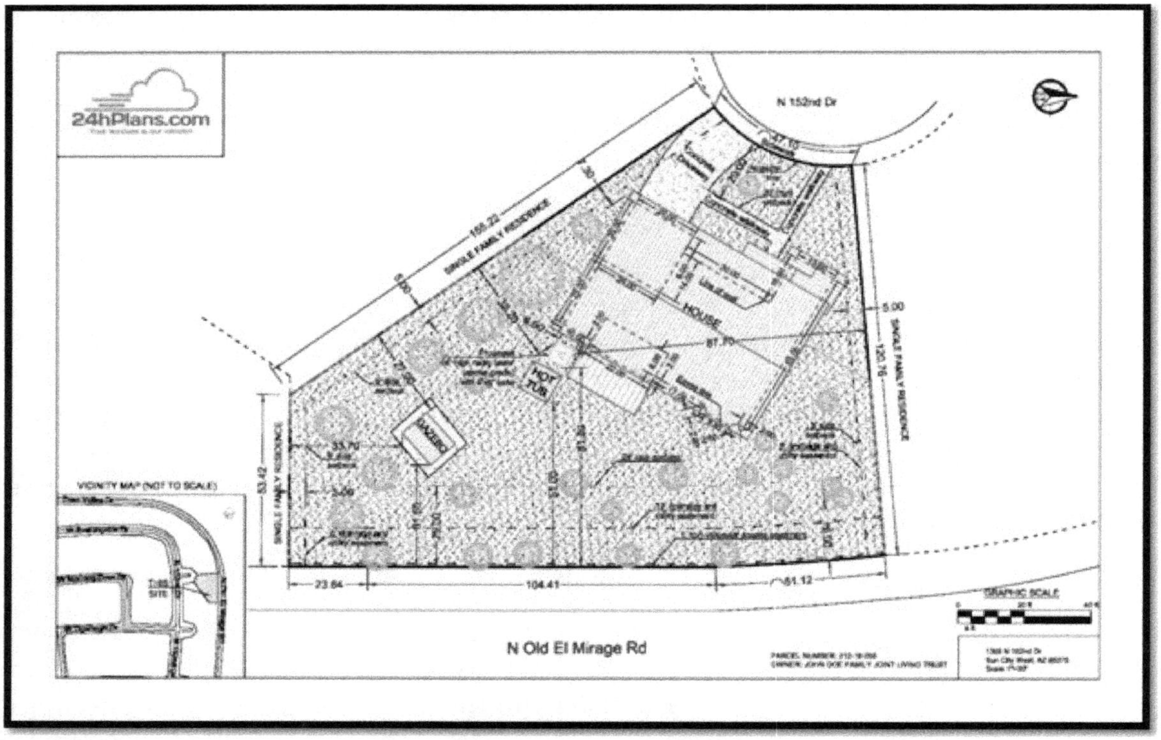

FIGURE 12.2 The residential plot plan places a specific house on a specific lot. All improvements, planned and existing, must be shown on the plan. Also shown are any recorded easements or setback lines.

The collective buyer desires, community requirements, and lot conditions must be organized to find a reasonable and mutually agreeable layout for the home, sidewalks, and driveway on the lot. Developing and getting everyone to agree on the plot plan is complex and must be done quickly for other things to progress in the starts process.

The plot plan is one of the first steps in the starts process. The agreed-upon final plot plan is usually signed by the buyer to avoid any difficult discussions once the home is built.

Plot Plan Development

From the community's preliminary or final plat, it is common practice for production builders to produce an empty plot plan template for each of the lots that they own or have under contract to buy. Each empty plot plan shows all the known easements and setbacks and other lot-specific constraints for that individual lot. The empty plot plans are blown up to a scale to fit an 8.5-by-11-inch page for better viewing.

Next, a floor plan template is made for each offered floor plan at the same scale as the empty plot plans template. When a home is sold, the salesperson arranges the buyer's floor plan template on the lot template to find the first best starting position for this house and lot combination.

This preliminary version may or may not be shown to and discussed with the buyer early on. Or, that early version may be sent directly to the architecture department for further development before it is returned to the salesperson for a formal meeting with the buyer. It simply depends on the flexibility and available options on that lot and whether the buyer has something already in mind. Small, tight lots do not offer much flexibility, so there will be pretty much only one place to locate the house on the lot.

Buyers usually accept a well-thought-out and well-presented preliminary plot plan from the salesperson. The buyer approval of the plot plan is an important step. The buyer must approve and sign the plot plan before the house may progress in the starts process.

Of course, the home must be built on the lot to exactly match the plot plan signed by the buyer. This must be a tightly controlled process because it is important to be sure that the plot plan satisfies everyone, including the buyers, the developer, and the city.

Plot plan development for spec homes, with no buyer participation, is much simpler and faster than starting a buyer's sold home. For spec homes, there is an internal process of developing and approving each plot plan. With these homes, the first plot plan is developed by the salesperson, further developed by the architectural group, and ultimately approved by other starts team members according to the prescribed process.

The importance of plot plan decision for a spec home cannot be overstated because a poorly placed home may hurt the marketability and sales price of that home if it's done without great planning and collaboration. Home placement is an important process as there is no redo when it comes to properly placing the home on the lot.

Job-Specific Architectural and Engineering Plans

Willingness to offer plan customization is often used as a company and home differentiator by homebuilders. Some buyers are willing to will pay more for their home when they can customize it. Buyers enjoy selecting from various plan options through which they can add bedrooms or entertainment rooms that are not part of the base house.

The spectrum of customization starts with buyers being allowed to select only from preset tile and flooring selections mounted on the garage wall, with absolutely no plan changes allowed at all. From there, it goes all the way up to sending every buyer to spend hours in a beautiful design center, sipping cappuccino with a design consultant and choosing from numerous plan options and myriad flooring, cabinets, and countertops and additional selections galore. The more extensive the options and selection offered, the more extensive and complex is the starts process.

No matter how much customization is allowed, the plans used in construction on the job site must reflect exactly the home you want built. Each company has its own proven process for revising the master plan set to reflect the buyer's selected plan options and other allowed changes. And do not forget that the sealed, city-approved set is what the city inspectors will use when inspecting your house. Therefore, all plan changes must be included on the final plan set submitted to the city and used on the job site when building the house.

Predesigned and Pre-Estimated Plan Options

The complexity of accepting unique plan changes from buyers and the magnitude of the problems resulting from getting changes wrong are so great that many builders do not allow many or even *any* plan changes, period. Other builders offer only pre-drawn, pre-estimated plan options. These options must profitable enough and selected often enough to be worth the builder's effort to pre-draw, pre-estimate, and bring to market on their marketing handouts and website.

The most time- and cost-efficient method of operation for a production homebuilder is to only build and sell the standard plan, offering no changes or options. This efficiency strategy only works in the sales office when the builder's retail prices are low enough to entice buyers to accept only the base house. Production builders offering at least some standard options to buyers will open their market to more discriminating buyers, higher prices, and more sales and profit.

Builders who choose to sell plan options need to design work processes around those offerings to be able to successfully get accurate starts packages produced for to each individual house start. As the complexity of the builder's offering increases, so, too, does the complexity of the starts processes.

Plan Set for Spec Homes

Setting up the plan options for a spec home, for which no buyer is participating in the process, is much simpler than starting a sold home. For spec homes, homebuilders have an internal process of proposing and approving the plan options and selections recommended by the salesperson who is preparing that initial worksheet on the proposed new start.

The spec home's first plan proposal is developed by the salesperson and then approved by other team members according to the prescribed method. Salespeople are the best people to know what floor plans they are selling and what options people are asking for and are likely to want on a future spec home.

The importance of making the right plan and option and color selection decisions for a spec home cannot be overstated. Think about it: If people are walking into the sales office all day and asking for five bedrooms and a media room, you'd best add the fifth bedroom and media room option to your next spec home if you want a quick sale. Adding or declining to add plan options is an important decision that affects the desirability of the home, its price, and its affordability.

There is a formal process of review and approval of the salesperson's requested selections and plan options, where the spec house package as proposed is justified as to how the proposed house is expected to sell at a price to support the proposed selections and options. The selections must add to the home's marketability and perceived value and therefore support themselves in the final sales price.

JOB-SPECIFIC BUDGETS AND SPECIFICATIONS

Production homebuilding software systems allow builders to build and maintain a plan master budget for each base floor plan and elevation they offer. Production builders are able to run the base house cost into a numbered job in their system. Once that base house cost is loaded inside the numbered job, the job now exists in the system and may then be adjusted and revised to match the buyers' selections and options and lot-unique costs, independent from and not affecting the plan master budget.

Depending on the building company's system, adjusting every start-related altered line item individually can take a lot of time. If a buyer selects wood floors where the plan says carpet, the carpet budget must be set at zero, and a wood budget must be added. Note that it is necessary to take this time and change every line-item budget for every affected work task. Remember that the objective is to send the start packages to the field with 100 percent accurate line-item budgets.

Job-Specific Construction Budgets

Construction budgets must be adjusted to reflect lot-unique budget changes sent in from the field as well as options and selections made by the buyer. The work of adjusting construction budgets is performed inside the purchasing department. In large builders, a team member is often dedicated completely to individually revising and producing the final budgets for each house start.

Selection Sheet

Project	Trace
Job	00627-BP13 204 Sage Meadows Drive
Plan	L-2367-F

Option	Cost Code	Change Order	Approved	Description	Qty	Attribute	Selection
Non Standard Option							
NonStan-01		001-015	1/21/2021	NSO Non-Standard # 1 Electric Fireplace & Surround. CHESMAR TO ADD 6" THICK, 8" DEEP, 6' WIDE MANTEL. (PAINT MANTEL PPG 1001-7 BLACK MAGIC).	1.000	Note	N/A
NonStan-02		001-016	1/21/2021	NSO Non-Standard # 2 Covered Patio - Ceramic Tile: Lv B Marazz Kaleido 8x8 (MOUD) with #93 Warm Grey grout	1.000	Note	N/A
NonStan-03		001-017	1/21/2021	NSO Non-Standard # 3 Kitchen Island - WRAP ISLAND SIDES AND UNDER COUNTER KNEE SPACE WITH 1/8" THICK BIRCH THEN ADD 1x3 TRIM OVER THE BIRCH SPACED TO LOOK LIKE BOARD & BATTEN STAIN WOOD AND TRIM TO MATCH FRONT DOOR	1.000	Note	N/A
NonStan-04		001-033	1/28/2021	NSO Non-Standard # 4 Front Exterior - (5) uplights. See exterior diagrams for details.	1.000	Note	N/A
NonStan-05		001-029	1/21/2021	NSO Non-Standard #5 Tile Full wall at Family (Fireplace) 90"x120" Lv B: Dal Tile Remedy 3x12, Herringbone lay, #02 Pewter grout	1.000	Note	See Detail Drawings
NonStan-06		001-030	1/28/2021	NSO Non-Standard # 6 Barn Doors - (2) at study. 1)Barn door at study entrance - provided by Chesmar to match interior door trim. 2) Study closet door - provided by November Design.	2.000	Note	See Detail Drawings
NonStan-07		001-031	1/21/2021	NSO Non-Standard #7 TRIM WORK: STUDY - Floating shelves. MBED - Tray ceiling. GAME - Wood panels. BED2 - Wall panels & trim See detail drawings.	1.000	Note	N/A
Appliances							
Applianc-S4		Standard		Appliances 1.0 Base Appliances w/ Range- Range JGBS66REKSS - Micro JVM3162RJSS - Dishwasher GDT635HS	1.000		
Bathroom Countertops							
Bathtop-01		Standard		Countertops - MASTER BATH- Base Contemporary Marble	1.000	Color	N/A See Upgrade
Bathtop-07		001-036	1/28/2021	Countertops - MSTR BATH- Granite / Solid Surface LVL 3	1.000	Color	Miami White - 3 CM Quartz
Bathtop-19		Standard		Countertops - Bath 2- Base Contemporary Marble	1.000	Color	N/A See Upgrade
Bathtop-23		001-040	1/28/2021	Countertops - BATH 2 - Granite LVL3 / Solid Surface	1.000	Color	Mianmi White - 3 CM Quartz
Cabinets							
Cabs-03		001-001	1/21/2021	Cabinets Level 3 w/ pulls & soft-close drawers	1.000	Door	Shaker
Cabs-03		001-001	1/21/2021	Cabinets Level 3 w/ pulls & soft-close drawers	1.000	Hardware	BMP4 - Matte Black

2/1/2021 10:48:40 **Page 1 of 5**

FIGURE 12.3 The selections summary, job-specific plans, and job budgets all must match. The starts package provides the builder with a detailed selections summary, such as the one seen here, of every option and finish item selected by either the buyer or the builder, in the case of a spec home start.

Only when the purchasing department is confident that the job-specific plans and budgets match each other is the job budget ready to finalize in the system and become the finalized start package. Finalizing the job permanently sets the budgets for that individual job, and any new budgets will be created as extras.

Job-Specific Specifications

The builder in the field receives job-specific plans and job-specific budgets, but the process must also provide the builder with job-specific specifications for each start.

These specifications usually arrive as part of the starts package in what will likely be called the *selections summary* or some such obvious name. This summary is very detailed and complete with manufacturer's model numbers and names and other shop drawings or whatever it takes so that the builder and subcontractor know exactly what materials in what design are to be installed.

The selections summary is the builder's guide used to confirm materials in advance of material orders and upon arrival with the subcontractor. The summary is used for reference and reconfirmation prior to delivery and installation. Experienced builders often reconfirm the selections summary sheet's version date to be sure they are using the latest version, especially if there is any question about the buyer's selection status.

OBTAINING GOVERNMENTAL PERMITS

The governing municipality and HOA will review your plans and possibly color selections and other specifications. The key is that you must build the home exactly as submitted. Therefore, your buyers must make all their decisions promptly before you can even approach or submit to the city or HOA for the required approvals.

Permit Plan Set

The plan set delivered to the city for a spec home or a sold home must reflect all selections and plan options selected for this address. No additional changes can occur once the plans have been submitted to the city. Also, that stamped and sealed set of approved plans that comes back from the city is the set that will be used for inspections in the field, so the house must be built to exactly match the permit set of plans.

Municipal and County Permitting

Governmental approval processes vary from municipality to municipality. Each governing entity enforces its own specific submission requirements. These are often extremely specific and must be followed exactly. The municipal websites usually provide their own checklist for building permit applications. Municipal plan reviews often require a plan review fee on top of the permit fee.

CHECKLIST FOR BUILDING PERMIT APPLICATION

	Items Required for Complete Submittal	Staff
☐	Completed Application for Building Permit	☐
☐	Plan Review Fee: 50% of Calculated Permit Fee (max $2,000) Technology Fee: $12	☐
☐	Approved or submitted Site Prep Permit SP/CSPP#________	☐
☐	One digital set of all items below uploaded to My Permit Now	☐
☐	COM Check / RES Check (Energy Calculations)	☐
☐	Asbestos Requirements (if applicable, commercial)	☐
☐	ADA Registration (if applicable, commercial) #EABPRJ ________	☐
☐	Architectural or Engineer Seal on AEI Occupancies (assemblies, educational, institutional)	☐

FIGURE 12.4 The permit checklist for San Marcos, Texas, is shown here.

Keep in mind that the plan set approved by the municipality is the one that must be built. Therefore, it is customary to submit plans to the city only after the plan set is approved by the buyer and HOA. However, given longer and longer lead times at the city, builders are more likely to apply for the city permit before the HOA has approved the plan set. Either way, the municipality plan set must be the final set.

Permitting Services

Obtaining the building permit is standard procedure for the starts group, which must become expert in the permit processes of all the municipalities in which they build homes. There are permit service firms whose only business is submitting, and processing plans through the city's permitting process. These firms can be extremely valuable to the builder. Permit service firms submit permits for your company and other companies and are in the permitting office every day. These firms have extensive experience in the city procedures and know how to process a permit most expeditiously. They review the plan sets before submission and may make suggestions that will help avoid a swing and a miss with your first submission. They know how to get the permit quickly and avoid delays.

These companies know exactly how plans get rejected or delayed as well as exactly what it takes to get the permit. If your company is new to a municipality, it may be advisable to seek professional assistance from a permit service to avoid making simple mistakes that can be very costly in delays.

HOA/COMMUNITY APPROVALS

Production builders usually build in new communities with robust community approval requirements and processes, as defined in builder guideline documents and agreed to in the lot purchase contract signed when the lots were purchased from the developer. Developers want some level of control over the houses built in their community, and that is accomplished by way of the builder guidelines. Builders must submit a plan set with all selections to the developer or the developer's architectural agent to review and ensure compliance with those guidelines.

The builder guidelines are often administered by the HOA, which often contracts with a professional architectural group to perform the formal plan review. The builders pay a plan review fee for each house to pay the cost of the developer's plan review service.

This plan review must conform to specific submission requirements. The approval submission includes job-specific plans and a summary of all exterior brick and paint selections. Generally, the community review is limited to the aesthetic exterior selections only. However, these guidelines also enforce home square footage limits and other plan option limits imposed by the developer in the guidelines. Each community is different, and each community's process must be followed.

Older Community HOAs and Deed Restrictions

CSM students who may someday build custom homes on vacant or tear-down lots in older existing communities must pay equal attention to the possible existence of builder guidelines or deed restrictions, which may also require a plan review by an HOA or architectural review agent. Contact the HOA for information and request any specific requirements for building a home in that community prior to designing or building anything.

Less obvious than the HOA requirements are the recorded deed restrictions. These follow the deed from owner to owner and are available at the courthouse. They may include specific clauses about what can and cannot be built. Usually, they have simple guidelines, such as no clotheslines or farm animals or no storage structures can be built. But do not bet on it being that easy; the restrictions sometimes require a specific type of roof, wall cladding like masonry, or lay out building setbacks that can have meaningful design implications.

There is a lot to consider when bidding new home projects in established, mature communities.

When the job-specific plans and selections are finalized, the HOA application may be started. This process can be long and frustrating, and experienced custom builders know to keep calm and keep the process moving forward.

POWER AND WATER METERS

Only with the city permit in hand is the builder able to obtain water and electrical meters. Both water and electricity will be needed promptly on the job site, so applications for both must be initiated immediately upon receiving the building permit and whatever other steps may be required locally to obtain the meters.

The electrician and plumber may the ones to obtain these meters, but the builder must open the billing accounts with the power company and for the water meter. Once the building permit is obtained, these meters are only administrative and should not be too challenging or likely to cause any delay.

SUMMARY

CSM students need to recognize the complexity of even reaching the point at which a home is ready to build in the field. The starts process is just as if not more complex in the other construction sectors, such as commercial and industrial. The starts process is a big part of the scope of work in any construction, and the process described in this chapter is somewhat applicable to permitting commercial or any other construction projects.

The starts process in homebuilding is distinct from other sectors in that this function is normally performed centrally by a team of starts specialists for all the company's home projects. CSM students may have noticed that the centralized starts process is like homebuilding's centralized purchasing process. Neither of these departments exist in commercial construction companies.

These centralized departments performing this front-end work needed to properly start a house is exactly the type of support that allows a builder to be a production builder.

REVIEW QUESTIONS

1. Why do production homebuilders usually have a separate starts group or starts department?
2. The builder in the field is responsible for which specific tasks in the starts process?
3. For what tasks is the corporate office responsible in the starts process?
4. Why is obtaining city permits and HOA approvals left till the end of the starts process?
5. What is usually included in the starts package that is delivered to the builder in the field?
6. Describe the plot plan development process used on a sold home start and the plot plan development process for a spec home start.
7. Describe the buyer selections summary and how it is used by the builder in building the home.
8. What is the goal of the plan review meeting with the buyer on a sold start?
9. How does a spec home get its selections and approvals for the starts process?

Figure Credits

CHAPTER 13

HOME DELIVERY PROCESS

INTRODUCTION

The end goal of the entire construction process is to deliver the beautifully completed home to the satisfied homeowners. The home delivery process is not just a financial transaction; it's also a transition to a new phase in the builder-client relationship. Delivering the home to the new owners is an emotional, legal, and financial transition that must be skillfully performed to achieve the desired outcome.

Production builders reach a point on every home at which they say, "This home is 100 percent complete." Now the builder must get the project manager, the company's licensed code inspector, the buyer's third-party inspector, the municipal inspectors, and the buyers themselves to say, "Yes! This house is 100 percent complete." Achieving universal agreement that the house is complete relies on the builder's organizational and communication skills as well as following the company's proven process of new home delivery.

Substantial Completion

Substantial completion is a legal term referring to an exact definition of what may be considered a completed project. The ability to say that a project is 100 percent complete and there is no work remaining is an important construction concept for Construction Science and Management (CSM) students in any sector to grasp. In commercial construction, this process is called *commissioning the project to the owner*. It is easy to know when you start a project, but it is hard to say when you are complete, especially regarding someone's home.

Home completion is difficult to define and enforce on homeowners. Good quality is subjective to every buyer, and a buyer is not inspecting the home to your company's specifications and allowed tolerances. Homebuyers inspect the home according to how the home condition is making them feel—confident or unconfident. The builder must control the process so that the home's good quality is demonstrated and is obvious to the buyer such that the buyer will logically arrive at the same conclusion: that the home is 100 percent complete.

The Process Will Determine the Outcome

CSM students must understand that the residential project delivery and commissioning process is significantly more subjective than is the delivery and commissioning of a commercial project. This intangible truth is an important distinction between residential and commercial construction.

Completion on commercial buildings is determined by inspecting engineers reading specifications. Completion on homes is determined by an amateur, nontechnical, emotional, and slightly nervous homebuyer.

Proper execution of the company's delivery process and the builder's interpersonal skills are just as important as the physical condition of the home. Experienced builders artfully take their clients through the twotothree-week delivery process with confidence, knowing that they will successfully lead their homebuyers through this final step in the construction process. The builder's organization and professionalism during these few weeks will give the buyers peace of mind, and they will calmly accept delivery of and be happy with their new home.

Process Overview

Although each community and municipality requires slightly different steps and a different sequence of this inspection and delivery process, here are the basics:

- Builder's final self-inspection and punchout
- Internal management walk
- Third-party inspections
  - Independent code inspector's final quality walk and signoff
  - Buyer's independent inspector (if the buyer has one)
  - City occupancy/final inspections
  - Energy inspection and program audit/certifications and infrared scan envelope inspection, architectural review committee and/or the homeowners' association inspection, and other locally specific inspections
- Buyer's first demonstration walk
- Buyer final walk

INSPECTIONS LEADING TO THE BUYER WALKS

The Management Walk

Once the builder has completed self- inspections and any associated punchout work, there is usually some form of an internal management walk by the project manager or another designated quality compliance manager.

The management walk is an internal walk designed to get one last set of company eyes on the home before inviting a series of outside inspections. The walking manager is usually the project manager but may be a warranty manager. The manager walks the home using the established final inspection checklist that the builder was presumably also using for the builder's own final inspections.

Project managers and their builders develop a method for these walks after delivering several homes together. Builders may even start the management walk by identifying items of likely concern, if any. Because everyone is familiar with the management walk process, these walks usually go well and have a predictably positive outcome. Upon completion of any resulting punchout from the management walk, the house is then ready for all the third-party inspections.

FIGURE 13.1 The builder walks the home with the project manager or another quality assurance manager with the company to review the general quality of the home by visually inspecting the home the same as the homebuyers will.

Independent Code Inspection

Production builders normally engage an independent code inspection (ICI) company to perform a predetermined set of inspections at specific stages of construction on all their homes. Each inspection in the series confirms and documents code compliance at each phase of construction. These inspections are an important component of the builder's quality compliance strategy. The final ICI is the last in a series of inspections that have followed the house from the start.

The goal of the final ICI is to ensure that the home is ready for the municipal final municipal occupancy inspection and the buyer's ICI report, if any.

The independent inspector's final inspection is an exhaustive and detailed list of specific items. These ICIs include literally hundreds, sometimes more than 500, inspection items covering every component and installation job on the home. This comprehensive inspection is designed to certify code compliance throughout the house and is for the builder's protection as much as it is for the buyer's protection.

Preparation is key for a successful final ICI. Meet the inspector on site and have the following items on hand:

- Buyer selection summary sheet
- Plot plan and a full set of plans
- All inspection pass letters or green tags
- Any house certifications: cable stress, insulation, energy program, etc.
- Any third-party inspection reports previously produced during construction
- Buyer correspondence indicating their special concerns, if any
- All remote controls for fans, garage door openers, fireplace controller, etc.
- All manufacturer's operation or warranty manuals

The independent code inspector's inspection results are often one of the measured items on the builder's bonus program. Experienced production builders pride themselves on how few items they average on all their final ICIs. A builder's cumulative inspection results are a valid indicator of the level of quality home that each builder is delivering. This is often a highly watched performance metric inside construction management teams.

Buyer's Third-Party Inspection

Buyers may or may not hire their own third-party code inspector. To avoid any delay, confirm with your buyers one or two weeks in advance if they are planning an ICI. If so, the builder needs to facilitate that appointment at the time the builder would like it to occur. Communicate early with the buyer to avoid stress and delay from a last-minute inspection.

The process for the buyer's inspection is the same as preparing for the company's ICI. Meet the inspector upon arrival and have relevant documents ready. The builder may already have the report from the internal ICI, and, if so, that report should be given to the buyer's inspector. If questions or issues arise during the inspection, call your company's code inspector for clarification and attempt to find an on-the-spot resolution to any raised issue.

Ideally, the builder collaborates with the buyer's inspector and gain that inspector's confidence, the home is in good condition, and the inspection results in a report written in a manner that is comforting to the buyer. Even if there are items found, the buyer will feel the tone of the buyer's inspection report, and that will play out in how well the buyer walks go.

In a case in which the buyer's inspector identifies negative or substantive items on the final report, the builder should promptly bring the company's independent inspector into the discussion. Let those two inspectors determine what is and is not a problem and decide on the appropriate resolution. Builders must avoid getting emotional or defensive in these cases. Look beyond the problem and at the resolution. Ideally, all issues can be resolved without involving the buyer.

City/Occupancy/Final Inspection

The house has already passed numerous city inspections and completed numerous ICIs. The city inspector likely already knows the house well and certainly knows the builder. For all these reasons, the outcome of the city's final inspection should be positive.

FIGURE 13.2 City inspectors inspect each phase of the home, and the finale inspection by the city should not bring any surprises because the inspector has already inspected the home several times.

Always meet the inspector at the house at the house and have all the documents from any prior inspections available. Follow the same attentive process with the city inspector as with the other inspectors. Never debate with city inspectors: instead, call your internal code inspector during the city inspection and allow those code technicians to arrive at an agreement or reasonable resolution.

The goal is to pass the inspection—and, if not, to agree to an action plan and reschedule the reinspection date. Maintain a positive tone and attitude, and there will be a positive result in the end.

THE BUYER'S FIRST WALK

The Demonstration Walk

The first walk is the moment to which the entire construction has been leading. The buyer's first walk is not an inspection walk; it is a demonstration walk.

The first walk is the commissioning of the home to the buyers by the builder. The builder will show and demonstrate the entire home to the buyers, including the home's high-quality construction and finish and demonstrating the operation and required maintenance for all installed mechanical, plumbing and electrical, irrigation and other systems in the home.

It is important to note that the first walk is the buyer's final opportunity to inspect the home for cosmetic issues. The home's warranty covers any failure of the home's systems or structural integrity, but the warranty does not cover cosmetic finishes noticed later that were present on the day of the first walk. Explaining this concept to the buyers the day of the walk takes finesse. Builders learn how to prepare the buyers for the walk, and its procedures and purpose.

The builder will set the tone and direction of the walk.

If the buyers are told they are to perform a quality inspection on their new home, they will do just that, hunting and finding every small defect that they can possibly identify. If told to, they will inspect the home. But this walk is a quality demonstration, not a quality inspection.

If the builder leads the meeting by explaining the commissioning, demonstrating, and onboarding purpose of today's walk the stage will be set for a great meeting. Start by explaining the series of inspections that have already occurred and passed and then present the completed home's quality in a positive light to the buyers and they will not walk the home like this is an inspection. The builder sets the tone of the meeting and thereby controls the outcome.

Builder is responsible for scheduling and conducting the demonstration walk.

1. **Schedule the walk time with the buyer at least seven days prior to the day you want the walk to occur and not fewer than 14 days before the anticipated closing date.** Ideally, set appointments for both the first walk and the final walk at this time. This gives you time to address all buyer items properly and have zero items remaining at the final walk. Allow ample time and be flexible to have a smooth and unrushed meeting.
2. **Arrive at the home early to the meeting and turn on all lights and confirm everything is perfectly clean.** Look in every room and closet and the garage. Walk the sidewalks and confirm everything is still perfectly clean.

3. **Have a written agenda for the walk and follow it completely** to ensure the entire home and all its components are is covered with the buyers. Review the agenda with the buyer to start the meeting. Use the agenda/checklist throughout the meeting to ensure the walk's efficiency and effectiveness in what can be a lengthy home demonstration and quality review.
4. **Builder controls the mood, pace, and sequence of the walk.** The walk is more about discovering their wonderful home than it is about finding hidden problems. The builder's people skills are key to a successful walk.
5. **Upon buyer arrival, meet the buyer in the driveway or at the entry.** Lead the group to the kitchen area to start the conversation about the agenda and goals of today's walk. In the kitchen, explain the walk process and all the paperwork associated with the walk that will need to be signed at the end of the walk. Everyone needs to understand the goals of the demonstration walk at the onset.
6. After the discussion in the kitchen, the builder directs the buyers back outside and **starts the walk outside the home at the curb. Walk and demonstrate the complete exterior first** and then proceed inside to do the same.
7. **The demonstration of the home and its components is an active demonstration.** Physically open and close or turn on and off whatever you are demonstrating. **Make the buyer open and close and operate whatever is being demonstrated. Do not simply explain how things work. Show the buyer.**
8. **At the conclusion of the walk,** the group should reconvene in the kitchen and agree on the final list of items, if any, to be addressed by the builder in the next seven days. **Buyer and builder both sign and date the appropriate forms**.
9. **Reschedule or confirm the appointment for the final acceptance walk,** ideally the day before the closing. Review the goal and limits of the final walk with the buyers. The goal and limit of the final walk is to only review the items from today's walk, not to make a new list.

The builder must stay positive and stay on the presentation points to accomplish the goals of the demonstration walk. To set proper expectations, experienced builders perform their new home demonstration by following a practiced presentation that they know will take the buyer through a complete, comfortable, and predictable demonstration walk.

No Home Is Perfect!

Home shoppers are usually looking at both new and existing homes. When shoppers walk existing homes and notice random, nonsymmetrical, oddly built things on that existing home, they think it gives the home character. In a new home, buyers do not want any of that kind of character. So, how is a builder to manage buyers who get fixated on one of these little "defects" in the new home? Below are a few strategies.

First, most production builders represent that the quality seen represented in the model is the quality buyers can expect in their home. Go look at the model home and see if you can find a comparable condition. Keep that in mind that the model home is the benchmark to which the home must be built, so if the model has such a defect, you can defend it or at least try.

Empty homes magnify the slightest imperfections. Put furniture in the rooms and hang a few pictures on the walls, and you will not be able to find the imperfection. See if you can find the same situation in the model and demonstrate for the buyer that this item identified usually occurs in all homes.

Second, apply what I call the *proportional solution*. Builders usually do what it takes to make the buyer happy in the end if there is a reasonable solution, one that is proportional to the problem. Small problems with small solutions, no problem. Small problem with big solutions—that is a problem.

For example, if a buyer notices that the brick layup has too many bricks of one color in a certain area (a brick mix issue). Is it reasonable to tear off all the brick from the house and rebrick it? No, that is not a reasonable solution. Some buyers do not care if the solution is reasonable or not. But for reasonable buyers, this approach can be effective. The builder can agree that this is a slight defect, but it is not large enough to tear off the brick. If the buyer insists, tell the buyer that the closing will be delayed a month, and the buyer will start to see the logic of a proportional solution to the problem.

FIVE KEY GOALS OF THE DEMONSTRATION WALK

1. Demonstrate and train the homeowner on the builder's warranty process and the included, and excluded, plan coverages. Covered items include mechanical, electrical, and plumbing; irrigation; yard lights; audiovisual options; and other systems in the home.
2. Demonstrate and train the homeowner on the operation and maintenance of each component of the home.
3. Documented all open items on the buyer's acceptance form that must be addressed by the builder during the next week, prior to the final acceptance walk.
   Note: This is the one and last time that the buyer may inspect for general cosmetics of the home. This gives the builder a clear view on what needs to be completed to satisfy the buyer.
4. This is a new phase in the builder/buyer relationship, and the buyer needs to understand how to get warranty service in the future.
5. The final goal of the first walk is to give the buyer peace of mind about their home and the warranty services they will receive in the future.

The Buyer's Demonstration Walk List

The documented cosmetic or repair items from the first walk, item 3 above, are the identified items from that walk and now become the builder's action items coming out of the demonstration walk.

Ideally, the buyer enjoyed the walk, feel good about their home, and have identified a short list of cosmetic or repair items that were documented on the buyer's acceptance form. This form reads that the buyer is satisfied completely with the home its included materials, and the workmanship on the home, save and except for the list of items identified on that acceptance form.

That list includes all the items that are remaining to be addressed in order for all parties to declare the home 100 percent complete. This is the home's last punchout list!

Now the builder knows exactly what must happen over the next week to be ready for the buyer's final walk. The builder must complete all items identified from the demonstration walk during the next seven days leading up to the final walk.

CHESMAR HOMES

Setting a Higher Standard

AFFIDAVIT OF ACCEPTANCE OF COMPLETION

To the Management of Chesmar Homes:

I have inspected my home at: ____________________
and hereby certify that it is complete. The quality of workmanship and material is acceptable to me. I understand that Chesmar Homes does not intend to do any further work of a cosmetic nature on my home. Any further work performed by Chesmar Homes will be limited to items of structural or mechanical nature. I further understand that the signing of this statement in no way relieves Chesmar Homes of its obligations, under warranties, concerning functioning of the equipment or any structural failures which may occur or become evident at a later date.

Please Note:
Signing this affidavit is in no way to be construed as a condition of this sale. It is a method to help the management of Chesmar Homes by ensuring your home has been completed.

*THIS IS YOUR TICKET TO CLOSING.
THIS AFFIDAVIT **MUST** BE PRESENTED TO THE TITLE COMPANY IN ORDER TO CLOSE.

Purchaser's Signature ____________ Co-Purchaser's Signature ____________

Print Name ____________ Print Name ____________

Date ____________ Date ____________

Construction Manager's Signature ____________ Date ____________

Print Name ____________

CM-002-11

Purchaser(s): ____________ Address: ____________ Date ____________

CHESMAR HOMES

Setting a Higher Standard

Buyer Orientation Meeting

The following items are common examples of cosmetic items not covered after closing.

We have taken special care to check each of these items for you. Please confirm that they meet your expectations as well.

- Scratches, chips or blemishes on appliances, cabinets, ceramic tile, countertops and marble
- Scratches, tears or discoloration of wood flooring or carpeting
- Scratches or chips on tubs, commodes, sinks or countertops
- Mismatched paint touch-up or hairline sheetrock cracks
- Scratches, chips, cracks and de-silvering of mirrors, windows and glass
- Tears and bent window screen frames
- Condition of landscaping, grass and trees
- Positive drainage away from foundation
- Settling of backfilled grade
- Condition of fence posts and pickets

Additional items needing attention prior to Buyer Orientation Final ...

Area	Description	Complete

Page 1

FIGURE 13.3 Production builders use a version of the buyer acceptance form seen here by Chesmar Homes to certify and document the buyer's acceptance of the home's materials, workmanship, and cosmetic condition. This multipage form memorializes the builder's presentation of the home's condition and demonstration of all the mechanical systems throughout the home. During the demonstration walk, the builder and buyer record any found items for attention during a buyer orientation meeting. There are the only items the builder works to resolve during the days following the orientation meeting and before the buyer's final walk.

THE BUYER'S FINAL ACCEPTANCE WALK

The final acceptance walk is the grand crescendo of homebuilding. During the final walk, the builder and buyer certify and sign that the home is complete and thereby formally move into a new phase of the builder-client relationship: the warranty phase.

Final Walk Objective

The one and only objective of the buyer's final walk is to demonstrate to the buyer that the items—and only those items—from the demonstration walk are repaired, replaced, or otherwise resolved and the home is now ready for the buyer to agree and sign off to that fact on the home's certification of completion.

Some items on a buyer's list may not be able to be accomplished during the week leading up to the final walk. For example, if a cabinet door needs to be replaced and it will take two weeks to get the door, this house will still close, but it will not be a zero-item home. A zero-item home is one in which all items on the demonstration walk were completed and the buyer signed off on the home being complete.

Closing homes with zero items is a big deal in the homebuilder world. Some builders pay an additional bonus for zero-item homes. Additionally, it is common for builders to be evaluated based on what percentage of their homes close with zero items. If one builder closed 24 homes with nine zero-item homes and another builder closed 24 homes with 21 being zero-item homes, you can guess who the better builder is.

If a home closes with an open item, the builder who built the home remains responsible for that home until any and all open items are complete, and the buyer sign the certificate of completion. It is always the builder's job to finish the home to the buyer's satisfaction is achieved, and the buyer ultimately signs the certificate of completion.

Difficult Buyers

A small percentage of your buyers will be difficult buyers. There is just no avoiding it. Sometimes we make buyers difficult through our builder errors, and sometimes buyers were difficult upon arrival. Managing difficult buyers is an essential skill for homebuilders.

SUGGESTIONS FOR CLOSING A DIFFICULT BUYER

1. Involve other managers on the buyer walks. You may need to double-team the buyer to be sure you can answer all questions during that walk and so that there are two people listening to and interpreting the buyer. The warranty manager is a good person to add to these walks because the warranty department is about to inherit this difficult buyer, and they might as well get to know each other at the walk.
2. Document everything that is said and agreed upon during the walk. Put your detailed notes in the system for future reference. New people performing warranty service for this home-owner will likely need as much context and background as possible to help them be successful on future warranty calls. Any advice you can give to co-workers about how to satisfy this homeowner is a good idea.
3. Revise the certificate of completion for this buyer's home to specifically address whatever is the open issue for the difficult buyer. Referring again to the example where the buyer does not like the brick layup, with too much of one color in one area, what do you do if this buyer is still upset the week of closing? That buyer may be clearly still upset about the brick. You may

have offered them all their money back and tell them it is up to them to accept or not accept the brick. They may say or be thinking, I will close today and come back and sue you later. In this case revise the certificate of completion to specifically read that the buyers were offered their money back and are not being forced to close on the home and, further, that they do accept the brick material and workmanship as installed on the home. If they will not sign that, then do not close. This is tough, but it is what you must do to end the discussion about that issue, the brick and avoid a lawsuit later.

SUMMARY

Homebuilding culminates in the completion and delivery of the new home to the happy new owner. Everything starting with purchase and development of the raw land into a beautiful community leading up to the construction and delivery of the new home has been leading to this home delivery moment.

The moment the keys are handed to the new homeowners was literally years in the making. The beautiful home and the happy owner are what homebuilding is all about.

REVIEW QUESTIONS

1. Describe the concept of substantial completion and why it is important in homebuilding.
2. Describe sequence, steps, and logic of the home delivery process.
3. Inspections by all the various third-party inspectors follow a similar form. What must the builder do to prepare for these inspections?
4. If code compliance issues come up during the city inspection or the buyer's independent inspections, what resource or support does the builder have to help navigate tricky technical issues?
5. What kind of walk is the buyer's first walk? What kind of walk is the buyer's first walk *not*?
6. What are the primary goals of the first buyer walk? What is the primary objective of the final buyer walk?
7. What are some strategies for dealing with a difficult buyer who is not happy or comfortable with a specific item or general condition?
8. What happens if there is a remaining open item from the buyer's first walk remaining incomplete at the time of the closing? How is completing open items managed, and who is responsible for completing the walk list and getting the buyer to sign off?
9. What are some strategies for dealing with difficult buyers?

Figure Credits

CHAPTER 14

THE NEW HOME WARRANTY

INTRODUCTION

Warranty is a separate process from the construction process and is the final chapter in the home-builder-client relationship. When effectively managed, the warranty process can enhance the client's complete homebuilding experience and build goodwill in the builder's local market. Homeowners who love their builder both before and after moving in refer their friends and family, which leads to future sales to new clients who are predisposed to trust their builder and love their new home. However, when poorly managed, the warranty process results in spiraling repair costs, legal expenses, and damaged builder brand image. Homebuyers become homeowners, and their relationship with their builder changes, along with the customer service processes in place to meet their needs from that point forward.

Goal of the Warranty Process

Creating happy homeowners is the primary goal for everyone in homebuilding, but it is most important for those in the warranty group. Achieving this goal is a complicated pursuit. Relative happiness can be measured by a survey, but the real evidence of a successful warranty group includes low legal expenses and high sales from direct homeowner referrals. Happy homeowners do not need lawyers and love to tell their friends and family about their new home. This wave of homeowner satisfaction creates the ideal environment for gaining market share and earning future high gross profit margins. Successful warranty management plays a huge role in the company's overall success.

Those in the warranty group sometimes feel like they are playing defense all the time; however, it does not have to be that way. With the happy homeowner as the goal, the group can develop an offensive approach to satisfying homeowners. Being proactive and responsive is a pleasant surprise for homeowners, reinforcing their positive purchase and construction experience. The result of great warranty service can be an almost evangelical level of builder support among the community's homeowners. During their weekend visit, home shoppers often ask the people they see around the community about their experience with their builder. A positive word from a homeowner will certainly result in a positive visit to the model home and possibly a new sale. Taking the offense on warranty can mobilize your homeowners as a stealth sales force in your neighborhood.

Turning the warranty group into an arm of the marketing and public relations team takes purposeful planning and a proactive warranty program, but the rewards are worth it. This positive and proactive culture also builds moral within the warranty group.

The Warranty Group

Large production builders usually have a separate warranty department, with a tenured professional as the warranty manager or director of warranty. Warranty managers are often promoted from inside the organization based on their personal history of customer-pleasing performance as a builder. The skills that made them great builders may or may not translate into great warranty managers. Successful builders have happy customers because those builders are great at avoiding problems. Managing the warranty department is about solving other builders' problems. Creative problem-solving and strong interpersonal skills are the keys to making a great warranty manager.

The warranty group usually consist of two teams: an office administration team and a field implementation team. The warranty administrators in the office receive, process, and monitor the numerous warranty requests received each week by the department. The warranty representatives working in the field personally evaluate the issues presented, schedule trades for the repairs, and, ultimately, manage each open warranty item to its terminal resolution.

Careers in Warranty Service

As discussed in the chapter on careers of this book, many entrants to the industry find their career start inside the warranty department. Construction Science and Management (CSM) students need to embrace that career starting point when offered. Warranty is a great place to start a career, learn the company's floor plans and specifications, and meet the company's trade partners. There is no doubt that many skills and a lot of knowledge are gained while working inside the warranty department.

Many of today's homebuilding company owners and large production builder executives started their career in a warranty department. You, too, can gain your footing in the field before taking on the responsibility of building houses.

The warranty department may also be a choice destination career position and not just a steppingstone role for aspiring builders. People who build their career inside the warranty group are highly valued, and many find this role challenging and rewarding. It takes a special person to work an entire career in warranty because the role can also be wearing after many years or decades.

The warranty department is strategic, nurturing the homeowner's goodwill earned during the buying and construction period and now extending and growing that positive experience after move-in. Warranty employees are the company's community goodwill ambassadors.

WARRANTY DECISION-MAKING

There are differences between warranty coverages provided by different builders, but the most important difference between warranty programs is in the warranty interpretation. Additionally, warranty plan interpretation is linked to company culture. Negotiating and navigating the homeowner's desires relative to the terms of builder's warranty coverage makes each warranty event a delicate matter.

During the first year of the warranty, builders accept responsibility for a wide variety of arising issues. Builders will act on almost any reasonable request during that first year. The exception to this rule is around home finish, cosmetics, or appearance-level warranty requests. Homeowners often discover "imperfections" after move-in and submit a warranty request for something to be reworked. This a path that the builder should not follow. The time for cosmetic repairs ends at the final walk. If the homeowner notices that the texture is a little too heavy on the one of the living room walls, never undertake a strip and retexture.

Warranty calls during the out years—years three through 10—are the service calls that most likely require experienced technical and business judgment. These out years of the warranty tend to reveal more serious problems that are more difficult to resolve to everyone's satisfaction. For example, in the event of a foundation problem, simply performing the needed structural repair is often not enough to dispose of the issue completely. The homeowners have reasonable concerns about the possible effects on the resale value of their home. That possible loss of value may even be specifically excluded from the builder's warranty and difficult to measure; however, the loss of value will still be in the discussion and disposition of the issue. All these important considerations and more, such as builder reputation, must be weighed in the warranty decision-making process.

Create the Best Warranty Program and Educate Everybody

Successful builders have a clear and proven protocol for delivering their warranty service. Great processes work best when everyone participating knows and understands the builder's warranty process, including the homeowners.

Getting the homeowner in step with builder's warranty program takes consistent and deliberate communication with the buyer leading up to the closing. Each preclosing meeting with the buyer is purposefully programmed to include information that sets the stage for the new phase of the builder-homeowner relationship. The verbal messaging and written communications leading up to the warranty period must be consistent across all builder team members.

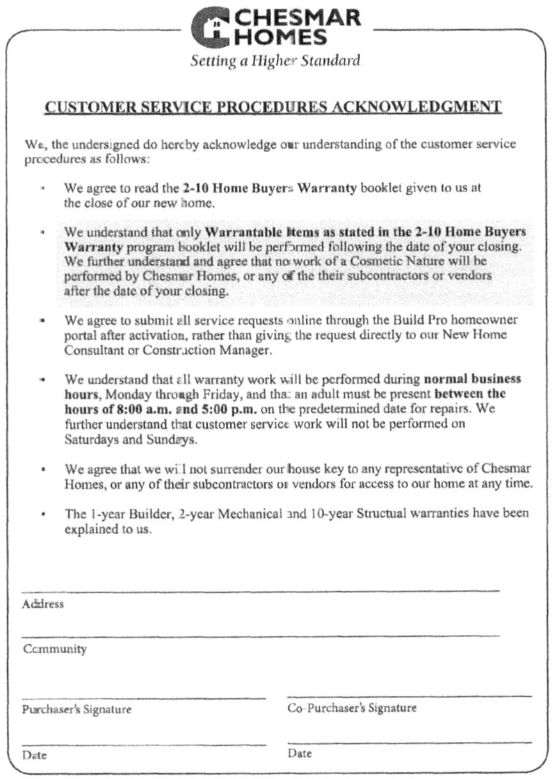

CHESMAR HOMES

Setting a Higher Standard

CUSTOMER SERVICE PROCEDURES ACKNOWLEDGMENT

We, the undersigned do hereby acknowledge our understanding of the customer service procedures as follows:

- We agree to read the **2-10 Home Buyers Warranty** booklet given to us at the close of our new home.
- We understand that only **Warrantable Items as stated in the 2-10 Home Buyers Warranty** program booklet will be performed following the date of your closing. We further understand and agree that no work of a Cosmetic Nature will be performed by Chesmar Homes, or any of the their subcontractors or vendors after the date of your closing.
- We agree to submit all service requests online through the Build Pro homeowner portal after activation, rather than giving the request directly to our New Home Consultant or Construction Manager.
- We understand that all warranty work will be performed during **normal business hours**, Monday through Friday, and that an adult must be present **between the hours of 8:00 a.m. and 5:00 p.m.** on the predetermined date for repairs. We further understand that customer service work will not be performed on Saturdays and Sundays.
- We agree that we will not surrender our house key to any representative of Chesmar Homes, or any of their subcontractors or vendors for access to our home at any time.
- The 1-year Builder, 2-year Mechanical and 10-year Structual warranties have been explained to us.

Address

Community

Purchaser's Signature | Co-Purchaser's Signature

Date | Date

FIGURE 14.1 Production builders educate buyers about their warranty process and procedures, often using their own version of an acknowledgment form as the one seen here to make sure buyers know what to do if they need service after closing.

Homeowner expectation setting begins with a well-educated and committed sales, construction, and warranty team who all know their company's protocol and are all communicating the same warranty message to their buyer clients. Expectation setting is a deliberate process.

Buyers must soon change their routine of communicating their home questions and concerns to their builder and salesperson and instead redirect all communication only to the warranty department. The builder/sales team has spent months spoiling their clients; now it is time to wean these buyers off that great relationship in the field and gently move them under the umbrella of the warranty department.

A clear view and understanding of the warranty process by homeowners and the warranty team alike will result to a smooth transition from the construction process into the warranty process. The move from construction customer to warranty customer is a relationship reset for the buyers, and builders must be clear about the new relationship methods of operation.

Warranty Program Administration: In-House or Outsourced

There is no best system of warranty administration. Each builder may choose the process that best matches their company's culture and goals. Today, many builders fully outsource their entire warranty service program to professional third-party warranty service providers. Other builders choose to manage warranty with their own warranty department employees. There are also hybrid programs in which the initial contact and most resolutions are managed by company employees, but the third-party warranty company takes over in certain and specified cases. All these methods can be successful at meeting customers' needs after closing.

The concept of outsourcing warranty is designed to isolate the warranty decisions to an objective provider who will approve or deny warranty requests based only on the merit of the claim relative to the objective terms and tolerances of the written limited warranty provided to the homeowner by their homebuilder. These professional, third-party warranty administrators are competent at evaluating issues and rendering objective decisions and empowered to implement the identified solution once communicated and accepted by the homeowner. Sounds simple and objective, right?

This professor is not a fan of the outsourced warranty program. Outsourced programs are highly successful for many builders, and the administrative advantages to this approach are not unnoticed or unappreciated by this professor.

It is important to deliver a culturally seamless experience to the homeowners by warranty employees who are cut from the same cloth as the other employees inside the company who delivered the happy homeowners. It is common for a homeowner's first warranty request to be managed by the original builder, the one person with firsthand knowledge of that home and that homeowner's concerns. The great culture that delivered the home to the happy homeowners can retain control of the process and deliver the same great service through the warranty period.

KEY INGREDIENTS TO SUCCESSFUL WARRANTY ADMINISTRATION

Risk Avoidance

Great warranty administration involves constant communication with the purchasing and construction groups to identify and avoid any repetitive problems from being built into future homes.

Risk Mitigation

Warranty management must weigh each issue and individually spend warranty dollars as needed to keep the warranty call under control and contained to as small a problem as possible. The risks we avoid are legal risks.

Finishing Homes

Yes, the condition of the homes entering the warrant program has a great deal to do with having a successful warranty program. Many builders include a step in their warranty program in which the warranty group walks and accepts every house into warranty. Whatever the process may be, completing homes 100 percent before closing and not sending preexisting issues to the warranty group is key.

Expectation Setting

Great builders have great warranty programs. If the homeowners understand the program and how they are to make warranty requests and then the builder responds with the needed service, the homeowners will be happy. If they are left to imagine for themselves how the warranty system works, they will certainly end up disappointed. Builders must educate their homeowners on the process of making warranty requests. Requests today are usually received through a homeowner portal, email, or regular mail.

Buyer Empathy

It is important to understand the stress and anxiety experienced by many homeowners when it comes to issues on their home. There is a genuine fear for many that something bad has happened to their home and that their builder might somehow abandon them. Keeping stress and anxiety levels low requires empathy-based action.

Anticipating Issues

The warranty program should also be proactive toward their homeowners. Most builders communicate with clients about impending weather conditions that might affect water pipes or heaters, for example. Builders often send reminders about normal home maintenance that homeowners should perform. Builders almost always perform a one-year service visit to the home to repair settling cracks or other expected cosmetic items that otherwise might upset some homeowners. Warranty managers anticipate and avoid issues.

A Good Written Warranty

The written limited warranty provided to homebuyers at the time of closing completely and clearly explains the exact coverages that the builder is covering. All warranty employees must be fully versed in and understand that written warranty. Builders often train their warranty employees on the interpretation of the most frequently occurring items to prepare them to deal with these specific items in a uniform way.

Subcontractor Collaboration

The builder's subcontractors are part of the warranty team. The subcontractor's responsibilities must be clearly spelled out in the subcontracts scope of work. Additionally, the subcontractors must be in sync culturally with the builder and work collaboratively and swiftly to resolve issues to the homeowner's satisfaction and in accordance with the limited warranty.

Sense of Urgency

Warranty administration is most successful when practiced expeditiously. The quicker and sooner that homeowners hear back and receive service from the builder, the better the chance is for a simple resolution and happy homeowner. Speed counts a great deal in warranty service!

Supervised Work

Homeowners are suspicious of workers left unsupervised in their home to perform a given task. It is not normal for builders to stand around and watch subcontractors perform their work. However, in warranty, it is sometimes necessary. The bottom line is that the warranty employee needs to at least appear to be present and inspecting ongoing work. This may be only once or twice during the work but must be more than zero. And when the workers are finished and leave the house, the homeowners expect the work to be inspected by the builder and their house to be clean. This is all hard to do, but when you do it, the homeowners will be very happy.

Good Reports and Alerts

The best systems provide for clear visibility of all the currently open warranty requests and historical performance metrics. In the warranty business, silence is not golden. Aging open requests must be carefully tracked, and homeowners must be communicated with throughout the duration of the open request. Warranty managers need clear visibility of and alerts on any developing issues and early enough that preventative measures may still be taken.

THE LEGAL SIDE OF THE WARRANTY

The Scope of the New Home Warranty

Builders offer their own limited warranty (the express warranty) to their homebuyers that cover specific components and workmanship for specific periods of one, two, and five years. Additionally, builders are legally obliged (implied warranty) to warrant most of the home's components and workmanship for the first year and must by law cover any structural defects for 10 years.

Between the builder's own express limited warranty and the implied warranties imposed by the state, builders have tremendous obligations to their buyers throughout years one, two, five, and 10. Additionally, the manufactured components of the home come with their own limited warranty from each manufacturer.

These various warranties have significant financial implications for the builder. For example, if a builder builds 500 homes a year, that means that the warranty department is managing 1,000 homes under the extensive two-year limited warranty and about 5,000 homes under the state's required 10-year structural warranty. The magnitude of the company's warranty obligation is significant and demands a professional approach at every step.

Manufacturers' Component Warranties

New homes are assembled from a long list of manufactured components, such as shingles, appliances, and flooring and heating, ventilation, and air conditioning (HVAC) compressors and plumbing fixtures. All these manufactured components carry their own warranties from the manufacturers. The homeowners must specifically accept all manufacturers warranties as part of their overall warranty on their new home. This provision provides that if there is a problem with the GE microwave, the homeowner agrees to deal directly with GE and not the builder.

Although the builder buys the component, the warranty provided by the manufacturer is for the benefit of the homeowner. The warranty team participates in almost all the warranty issues associated with a manufacturer's products, but ideally, the issue is resolved to the buyer's satisfaction without too much involvement by the builder.

The Builder's Express Limited Warranty

An express warranty is a written warranty provided by the homebuilder to their homebuyers. The builder writes the express warranty to expressly state exact coverages on each element of the home for specific time periods. The express warranty is and written by the builder and is designed to match exactly the builder's warranty program. All builders offer some form of an express warranty, although doing so is not required by Texas law.

TYPICAL LIMITED WARRANTY PROGRAM COVERAGE

- One- or two-year coverage on all workmanship and exterior materials
- Two to five years on windows, HVAC, plumbing, and electrical
- Ten years standard coverage for major structural defects

The scope of the coverage provided by each builder's limited warranty usually covers one or two years on almost everything on the home; two years and five years on various manufactured components and critical systems, such as the roof; and 10 years on the home's structural integrity.

Implied Warranties

Two Implied Warranties Provided to Homebuyers by Texas Law

Both implied warranties provide homeowners with "a reasonable time" to discover and report a defect.

Implied Warranty of Good and Workmanlike Manner

This warranty guarantees the home is built with quality materials and built at a level of skill equivalent to that which would be provided by any reasonably proficient builder doing the same work.

Although this is somewhat subjective, *workmanship* is usually defined in narrative terms of specific tolerances and allowed workmanship.

Implied Warranty of Habitability

This warranty guarantees that the house is suitable for human habitation and sufficiently safe and sanitary, stating that homeowners must be able to live in the home without threat to their health or well-being. This is an extremely low bar for home performance and does not hold up well in legal proceedings.

OTHER LEGAL CLAIMS

Homebuilders may also find themselves in legal difficulties with homeowners for items linked to the sale of the home and any representations about the home or the lot made by the builder to the buyer. These claims are of a different nature than those regarding the performance or workmanship of the home.

Claims of Breach of Contract

Breach of contract is a claim of a demonstratable difference between the home promised at the time of the sale and the home delivered, asserting a promise made and a promise broken. Production builders have great sales contracts and normally protect themselves well from these claims. CSM students need only be aware of the nature of this claim and always be confident when answering buyer questions and not be afraid to say, "I don't know, but I will get an answer."

Claims of Negligence

A claim of negligence accuses the builder of failure to exercise standard and normal care, which then resulted in a specific construction defect. Failure to build to current legally imposed building codes is the best example of negligence. Homeowners need to demonstrate and quantify economic damages suffered to collect damage payments. Claims of negligence are why builders place so much emphasis on third-party inspections throughout the construction process and inspection record retention.

TEXAS RESIDENTIAL CONSTRUCTION LIABILITY ACT (RCLA)

When the builder's warranty process fails to satisfy the homeowners, they may turn to an attorney. The Texas RCLA provides builders and homeowners with a formal process for making a claim of a construction defect and a process for responding to the home defect claim received by the builder. The definition of a *construction defect* is also described in the legislation.

The process spells out specific notice requirements by the homeowner to the builder and specific response timelines required of the builder. Both parties must follow the process and timelines carefully to protect the rights of both parties.

RCLA Process

The homeowner must give the builder notice of the defect using certified mail and return receipt and provide reasonable details and photography of the defect to the builder. **This certified notice starts the timeline in this claim process.**

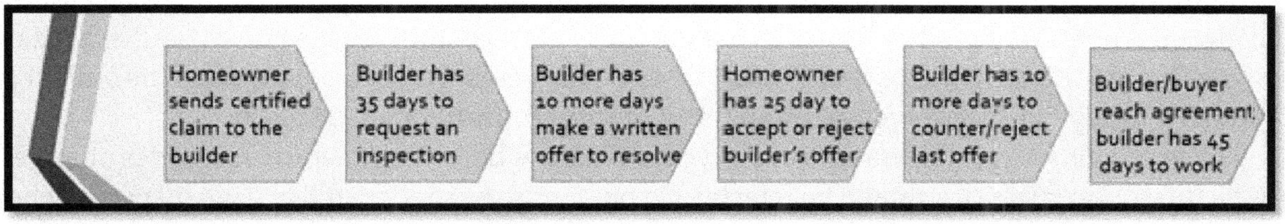

FIGURE 14.2 The Texas RCLA provides protection and process for homeowners and builders to follow when issues become legal.

The builder has a maximum of 35 days to respond by requesting to inspect the home and the identified defect. This is an especially important provision of the RCLA: The builder has the right to inspect the defect prior to any modification or repair by the homeowner. The homeowner cannot hire a third party to perform a repair on the defect and send the builder the bill. The builder has the right to inspect the defect as found and perform the repair themselves.

The builder must reply with written offer of repair or settlement within 45 days of receiving the initial claim. A timely offer of a reasonable remedy is important in keeping this process moving.

The homeowner then has 25 days from receiving the builder's offer to accept or reject the offer. If no reply is made, the offer is considered rejected. If the homeowner rejects the builder's offer, the homeowner may choose to counter the offer or may advise the builder as to why the offer was rejected. The builder is then given 10 days from the homeowner's reply to revise the repair or payment offer or reject the homeowner's counteroffer. Often the quickest and most satisfying way to resolve these heavy issues is to give the homeowner a check and let them perform the repair to their own satisfaction.

Following the RCLA rules is a legal process and requires legal assistance. Never try and navigate the RCLA process without the guiding hand of your attorney!

Right to Inspect and Repair

The principal advantage to the builder of the RCLA process is the opportunity to inspect and repair. Homeowners would otherwise hire their own subcontractors to perform who knows what work and then send the builder the invoice. Builders deserve this protection and the right to perform the needed repairs themselves.

Builders have warranty agreements with the subcontractors who originally performed the work, and the builder and subcontractor have the right to repair their own problems. No one would think about buying a new car, performing a repair at a local repair shop, and then taking the invoice to their new car dealer and reasonably expecting that dealer to reimburse them for the repair. The same holds true for new homes.

THIRD-PARTY HOME INSPECTIONS

First, it is common that a third-party code inspector, sometimes called an *independent inspector*, is hired by a homebuyer to perform a home inspection on the new home. Next, the builder's homeowners become home sellers, and their buyers may also hire another third-party inspector to inspect the home and produce a report. Should that report contain any negative comments that trigger the second buyer to request repairs, those repair items are often presented to the builder's warranty group. This is everyday stuff in the warranty administration business.

These second-buyer inspections occur years after the home completion and codes change. Code infraction questions should be referred to your company's independent code inspector who performed the inspections at the time of construction. Almost all not-to-code items are build-date sensitive, meaning the home was originally built to the code in place at the year of the build but perhaps not to today's code. A letter or information from the builder or builder's inspector usually ends these issues without further discussion. If not, let the two inspectors propose a resolution.

These inspections may also report on the general performance of the home, small things that failed but probably should not have. Here is a suggestion: When there is a solution that is doable and not very expensive, implement it quickly. Even seemingly larger or difficult items, such as a sagging roof area, may have a quick bracing solution. These items are common and will only be resolved in-person by a company warranty representative. Some items are matters of opinion and can be discussed and resolved. However, the best suggestion is to add additional bracing or whatever you can do to quickly resolve the issue. In and out and the problem solved is the best approach to items that can be repaired quickly and inexpensively.

KNOW YOUR COMPANY'S COVERAGES

Working in the warranty group can be rewarding technically educational. The key to successfully performing the work on behalf of your employer and the homeowner's is to be thoroughly versed in the exact terms, tolerances, exclusions, specific coverages, and requirements. Equally important to understanding your company's coverages is an understanding of the company's intention, desired approach, and likely interpretation in these challenging warranty issues.

Executives and project managers and the warranty team members must all be on the same page of the same warranty book. Everyone must be confident in their agreement because homeowners inevitably call up the ladder, wanting the next manager, whenever they get an answer they don't like. "No, that is not covered in the warranty, and we will not fix that item." The warranty group must know that their management has their back on interpreting and implementing the warranty terms. The entire process of the homeowners interacting with several employees on the same issue creates an opportunity for cross-communications. Everyone must be well trained to hold back on comments until the entire team is on the same page.

When everyone understands both what the warranty says and collectively how we want difficult issues managed and then all employees communicate with one voice, then all these warranty challenges resolve themselves in time.

SUMMARY

The warranty process and continuing warranty obligation are often overlooked when looking at the home construction process and the residential industry at large. However, the warranty obligation looms heavy over all decisions successful builders make, from their specifications to their choice of subcontractors.

The warranty department is a strategic force inside the homebuilding company. Inside these huge companies, those groups are proactive problem preventors. They are the tip of spear in a process of continuous improvement and are constantly delivering performance feedback to the construction and purchasing teams.

The warranty department can also be a strategic part of the sales force. The builder has worked for most of a year to build a beautiful home and a satisfied homeowner, so it is incumbent on the warranty team to build on that successful relationship to complete the entire homeowner satisfaction cycle.

REVIEW QUESTIONS

1. Describe the meaning of the *implied* and *express limited warranties*. What are the normal coverages of each warranty?
2. What are the key features of the Texas RCLA, and how does that play out for the builder and the homeowner with a home problem?
3. What are the pluses and minuses of outsourcing the warranty process or retaining that process inhouse?
4. Discuss the key ingredients to successful warranty administration or the items that stand out most to you.
5. Discuss the importance of all company employees, both in the field and in the office, being on the same page regarding warranty coverages and interpretations.
6. What are some of the advantages of starting one's career in the warranty group?

Figure Credit

CHAPTER 15

THE HOMEBUILDERS ASSOCIATIONS

INTRODUCTION

The National Association of Home Builders (NAHB) was founded more than 80 years ago to advocate for the nation's homebuilders, remodelers, developers, contractors, and associated businesses. The NAHB is a federation comprised of more than 700 state and local homebuilders' associations and more than 140,000 members across the nation. NAHB members build approximately 80 percent of new homes constructed in the United States.

The NAHB strives to protect the American dream of housing for all while working to achieve professional success for its members, who build communities, create jobs, and strengthen our economy. The NAHB is a trade association supported by businesses that operate in the homebuilding (single-family and multifamily), remodeling, and land development industry.

Image 15.1

When members join the NAHB federation, they do so at the local homebuilders' association (HBA) level and automatically become a member of the local HBA, their state association, and the NAHB. There are three classifications of members: builder, associate (supplier), and affiliate (an affiliate member can join at a lesser fee if another individual within their company holds a full builder or associate membership).

Image 15.2

THE TEXAS ASSOCIATION OF BUILDERS

Founded in 1946, the Texas Association of Builders (TAB) is an affiliate of the NAHB and has 26 local home builders' association and nearly 10,000 members across Texas. Representing **almost 750,000 jobs** and approximately **$70 billion** annually in the Texas economy (according to 2022 numbers), TAB members and the state and local associations play a critical role in providing housing for Texans. The TAB leadership and staff are always available to members to learn how to enhance the membership experience, help find business solutions, and provide a great value for a member's dues commitment.

How We Support the Homebuilding Industry

The TAB's advocacy efforts may be the most important service that the association provides. Nearly 25 percent of the cost of a new home is due to legislative and regulatory mandates. Without the efforts of the TAB and its members in educating and working with statewide and locally elected officials and regulatory agencies, the cost of housing would skyrocket, making the dream of homeownership one that most individuals and families could never achieve. According to the Texas Real Estate Research Center at Texas A&M University, for every $1,000 increase in the price of a new home, 22,000 households are priced out of the homebuying market.

That's why it's important that those who work in the homebuilding industry join their local homebuilders' association, which allows us to work together for the industry to ensure that homebuilding is not unduly burdened with taxes, fees, or regulations so that we can help to keep housing affordable for everyone.

"Every man owes a part of his time and money to the business or industry in which he is engaged. No man has a moral right to withhold his support from an organization that is striving to improve conditions within his sphere." —Theodore Roosevelt

Having a strong, effective government relations program is critical to the success of association members and their businesses. TAB's advocacy efforts at the state level are estimated to save $15,000 per home start. TAB's political action committee, HOMEPAC, is an essential part of TAB's advocacy program and maximizes political contributions by association members by supporting pro-housing candidates and incumbents for both statewide office and the Texas legislature. It is through voluntary contributions that HOMEPAC can help preserve a positive political climate for the residential construction and development industry in Texas. Without the HOMEPAC's participation, decisions affecting the future of the residential building industry will be influenced by other industries and groups, many of whom represent interests directly opposed to ours.

HOMEPAC
TEXAS ASSOCIATION OF BUILDERS

Image 15.3

TAB's Rally Day occurs every other year when the Texas legislature is in session. This is a great opportunity for members' voices to be heard when they meet with their elected state senators and representatives to talk with them and educate them on the homebuilding industry and TAB's legislative agenda. In addition, TAB's Voter Voice gives members immediate access to decision-makers to support the TAB's government relations efforts and enables TAB's legislative team to implement a coordinated communication and grassroots advocacy campaign through email, e-newsletters, and action alerts. The TAB's government relations team, HOMEPAC, Rally Day, and Voter Voice provide the residential building industry in Texas with a comprehensive advocacy effort—one that is effective in positive policymaking for our industry.

How We Help Your Business

Membership in homebuilders' associations facilitates building relationships and community involvement and provides members with credibility and a competitive edge by demonstrating to the consumer that they are part of a professional organization. The exchange of ideas, knowledge, and solutions to challenges between professionals in the same field cannot be underrated. Member can accelerate their career by taking advantage of the many educational and networking opportunities available to members—opportunities that are not accessible to those outside of the HBA federation.

Another benefit of membership in the association is access to programs and services that are not available to the general public. A TAB builder member can save on average thousands of dollars a year by taking advantage of the TAB's various discount and rebate programs. TAB's Texas Residential Construction Contracts Package, available only to builder members, can protect your business while saving you potentially thousands of dollars in legal fees. TAB has a comprehensive insurance program that includes builder's risk, general liability, small-group benefits, and personal lines of insurance such as health and home/auto. These are only a few of the exceptional services developed exclusively for TAB members.

Throughout the year, TAB offers educational webinars, and members automatically receive local HBA and TAB print and e-publications, including *Texas Builder* magazine, the TAB News Briefs, and TABloid and Legislative Line e-newsletters. Members also receive the NAHB's *Builder* magazine and may sign up to receive the NAHB's e-newsletters and blogs at nahb.org.

How to Become an Involved Member

Your local HBA, TAB, and NAHB hold committee and board meetings, conferences, and networking events throughout the year. Volunteer to help with a networking event. If you are interested in the governance of your association or are looking for a way to get to know your peers who share similar interests, join a committee and attend association-hosted conferences. Most committee meetings are open to all members, so feel free to attend any meeting that appeals to you.

Owned and operated by the TAB, the Sunbelt Builders' Show and Conference is designed to help builders, remodelers, and associated companies grow their businesses. The two-day show is the largest in Texas and gathers more than 200 top manufacturers and suppliers each year to showcase the best new products, services, and technologies to thousands of builders, remodelers, and developers. The show floor also hosts quality education and demonstration sessions for attendees. Because this is not a consumer show, builders can become educated on the latest products while building relationships with industry-specific suppliers.

SUNBELT BUILDERS SHOW™

Image 15.4

The conference, held in conjunction with TAB's summer committee and board meetings, features keynote speakers and several networking events. The TAB also holds committee and board meetings in the fall and winter, allowing you additional opportunities to get to know your peers from across the state.

Image 15.5

Local home builder's association also have many opportunities in which you can become an involved member. Attend meetings and events and volunteer to serve on committees. Be a part of the HBA's parade of homes and/or home and garden show. The more you give to your association, the more it will give back to you.

SOLVING THE WORKFORCE SHORTAGE AND CHARITABLE ACTIVITIES

The Texas Builders Foundation, the charitable arm of the Texas Association of Builders, serves as a resource to students, the homebuilding industry, and the citizens of the state of Texas. Its mission is to support and develop the future of Texas' construction workforce and facilitate the charitable activities of the residential construction industry. The foundation awards scholarships to students enrolled in construction programs in Texas' technical and trade schools, colleges, and universities. Additionally, the foundation is a source of information and assistance to organizations or institutions that need support in starting or enhancing construction trades programs in their area's high schools and community colleges. The Texas Builders Foundation is a 501(c)(3) charitable foundation, and all contributions to the foundation are tax-deductible to the extent allowed by law.

Many local HBAs have workforce development committees that work with their local high schools and colleges to promote the residential construction industry as a professional career choice, and mentors for students are in demand. If you are interested in supporting charitable activities, the HBAs participate in a variety of opportunities to serve your community.

TAB Values Its Members

The Texas Association of Builders has a variety of programs to honor individuals or companies that have provided excellent service or made significant contributions to the association and our industry. These awards are highly regarded throughout the industry and, much like membership in the association, proves to consumers a strong commitment to excellence in homebuilding by member nominees and winners. Award programs include the Star Awards, TAB's Excellence Under 45 Awards, the Texas Housing Hall of Honor, Of-the-Year awards, and Presidential Distinguished Service Awards.

Your local home builders' association also hosts awards programs and is proud to showcase and reward the talented individuals and exceptional homebuilding companies in their communities. These awards are not only important to your company, but they also advertise the outstanding work of the professionals within the residential construction industry.

Image 15.6

For more information on the Texas Association of Builders or to find the local home builders' association in your area, visit TexasBuilders.org or call (512) 476-6346.

NAHB STUDENT CHAPTERS

Every day, high school and college students in construction programs are learning skills to prepare for the exciting opportunities within the building industry. The NAHB Student Chapter program is dedicated to enriching students' educational experience and offers firsthand exposure to the real world of homebuilding and the industry through membership and education and networking opportunities. NAHB Student Chapter members also have access to the same member benefits as a builder, associate, or affiliate member. Each NAHB Student Chapter is chartered by the NAHB board of directors through a local home builders' association , and members of the chapter are encouraged to become active in the local HBA.

NAHB Student Chapters also runs the NAHB Student Competition, a real-world construction management and planning contest for students at the high school (dues are $5 per student, or $200 for unlimited secondary student members), two-year college, and four-year college/university levels (dues are $25 per student, or $400 for unlimited postsecondary student members). The competition is held every year at NAHB's International Builders' Show. For more information about the Student Chapter program, visit nahb.org/studentchapters.

Contact Information

National Association of Home Builders
WATCH AT https://tinyurl.com/duf6jtac

Texas Association of Builders
WATCH AT https://tinyurl.com/3kmhpnuz

Texas Local HBAs
WATCH AT https://tinyurl.com/52hfbjyw

HOMEPAC
WATCH AT https://tinyurl.com/yckhdpv7

Sunbelt Builders' Show
WATCH AT https://tinyurl.com/r5b82kw2

Texas Builders Foundation
WATCH AT https://tinyurl.com/2fj67wfk

NAHB Student Chapters
WATCH AT https://tinyurl.com/277caj69

Figure Credits

IMG 15.1: Copyright © by National Association of Home Builders. Reprinted with permission.
IMG 15.2: Copyright © by Texas Association Of Builders.
IMG 15.3: Copyright © by Texas Association of Builders. Reprinted with permission.
IMG 15.4: Copyright © by Sunbelt Builders Show.
IMG 15.5: Copyright © by Texas Association Of Builders.
IMG 15.6: Copyright © by National Association of Home Builders. Reprinted with permission.

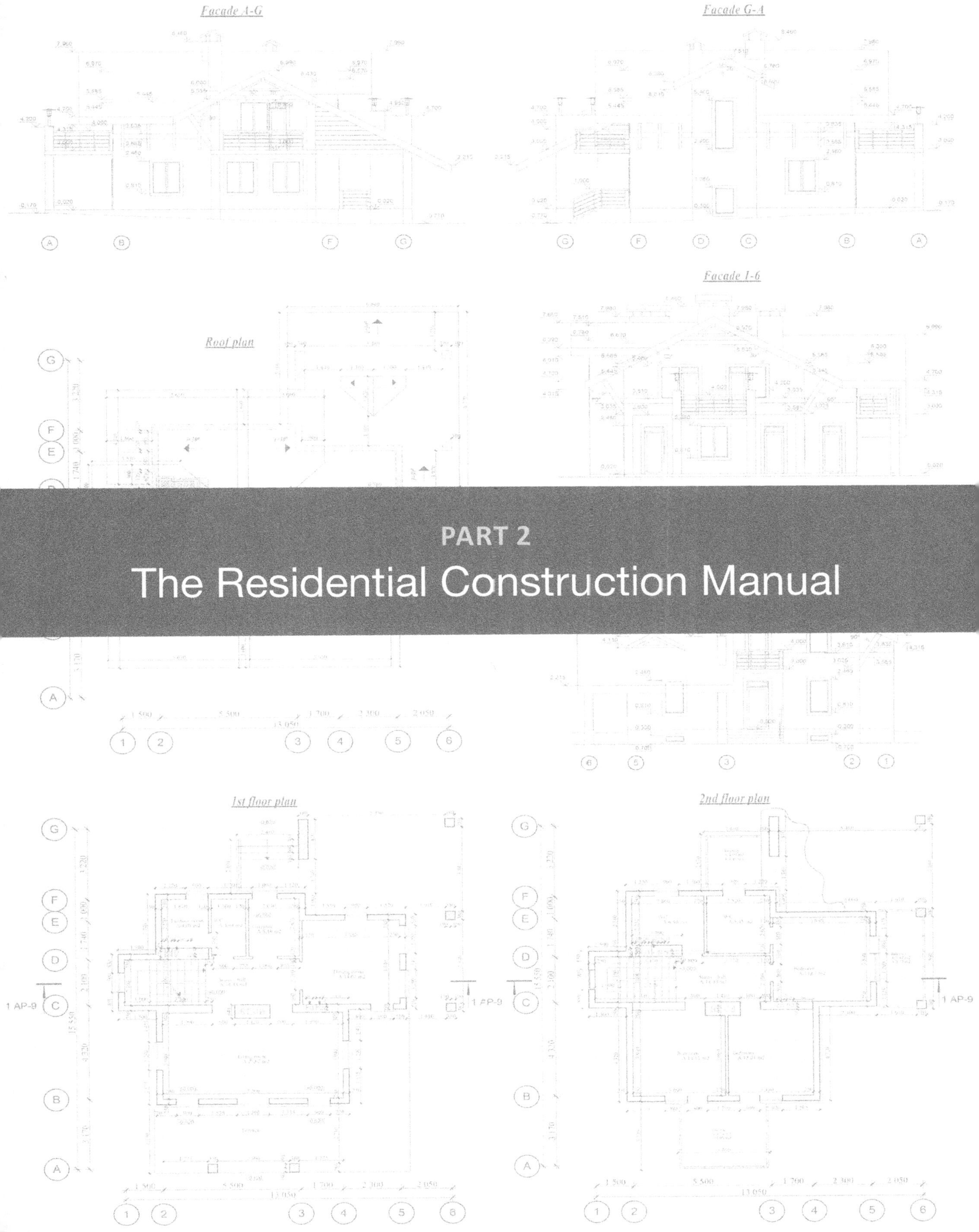

PART 2
The Residential Construction Manual

RESIDENTIAL CONSTRUCTION MANUAL INTRODUCTION

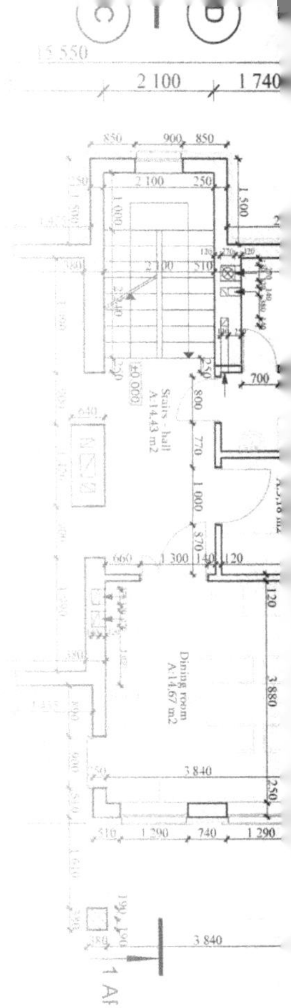

This construction manual is designed to introduce and instruct college Construction Science and Management students on the production homebuilding sequence and tasks and the responsibilities of the builder during the new home construction process.

This construction manual teaches the stages of construction following a proven method of instruction used for more than 50 years by production homebuilders to train new employee builders. The late, great General Homes introduced one of the earliest of the production construction manuals, from which this method of stage-by-stage construction instruction became the homebuilding standard teaching tool. This basic manual is designed to extend that homebuilding training tool into the college classroom.

The student learning objectives designed into this manual include learning:

- **The staged process and sequence of new home construction**
- **The general scopes of work for each trade at each task**
- **The builder's quality confirmation points at each task**
- **The principles of lead time and forward scheduling**
- **About the starts process and the home delivery process**

The stages and tasks included inside each stage herein are illustrative of the stage and task concept used in production homebuilding. The number of stages and stage task inclusions vary significantly from one builder to another.

This manual guides the student through the construction of a basic, low-feature-level home being built in a high-production environment. Because of the constraints of the semester teaching cycle, this manual is not designed to cover every material and installation method type common to homebuilding. The manual is designed to introduce the student to home construction basic sequencing and prepare the construction graduate to quickly understand what to expect and be able to contribute immediately upon graduation and starting a homebuilding career.

This manual's sequence, specifications, and tolerances are only illustrative of what students will encounter when building a home. Building methods and materials are always evolving and are done differently in different cities by different builders, and therefore this manual is never complete, current, or exactly accurate. This construction manual is designed for collegiate construction science and management instruction and not for use in the actual construction of a home.

NOTE FROM THE AUTHOR/EDITOR

Practicing homebuilders are welcome to submit stage or tasks edits or new task pages and/or other revisions to this residential construction manual directly to me at fwh10@ txstate.edu. This first edition is just the starting point for college-level homebuilding instruction, and all improvements or additions are welcome for future editions.

I was not the most qualified homebuilder to write this manual; I was just the one who had the time and opportunity. The chapters of *The Production Homebuilder* and this construction manual were written to provide our homebuilding industry with a skeleton textbook on which we can build new and improved college homebuilding curriculum going forward.

General Homes

CONSTRUCTION

and

CUSTOMER SERVICE

MANUAL

GENERAL HOMES

HOUSTON REGION

Image C.1

General Homes Manual Cover

Version dated August 1, 1982

THE RESIDENTIAL CONSTRUCTION PROCESS

Stage 0: The Starts Process

Stage 1: Lot Preparation Stage

Stage 2: Slab Stage

Stage 3: Rough Grade Stage

Stage 4: Frame Stage

Stage 5: Cornice and Shingles Stage

Stage 6: MEP Rough-In Stage

Stage 7: Drywall Stage

Stage 8: Interior Trim and Cabinets Stage

Stage 9: Final Grade Stage

Stage 10: MEP Trim Stage

Stage 11: Inspection-Ready Stage

Stage 12: Signoffs and Buyer Walks Stage

Figure Credit

ALL STAGE EXAM SOLUTION

STAGE 0	STAGE 1	STAGE 2	STAGE 3	STAGE 4	STAGE 5	STAGE 6
Starts Process	**Lot Preparation**	**Slab Stage**	**Rough Grade Stage**	**Frame Stage**	**Cornice & Shingles**	**MEP Rough-In**
8	**5**	**6**	**3**	**2**	**10**	**8**
Buyer Selections	Tree Protection	Set Forms	Release Frame Package	Chalk out the Slab	Measure for Cabinets	Plumbing Top-Out
Home Site Evaluation	Vegitation/Tree Removal	Plumbing Grounds	Stress Cables	Frame the House	Termite Treatment	HVAC Rough
Job Unique Budgets	Build House Pad	Makeup & Float	Lot Clean & Rough Grade		Frame Inspections	Electrical Rough
Job Unique Plans	Pad Certification	Pre-Pour Inspections			Sheathing Install	Electric Underground
Locate Corner Pins	Cut the Curb	Pour Foundation			Cornice Install	Low Voltage Rough
Set T-Pole		Strip & Dress Slab			Roof Decking Install	Vacuum Frame
Install Erosion Control					Exterior Paint Prime	Pre-Cover Inspections
Permits Received					Set Fireplace	Buyer Pre-Cover Walk
					Shingles	
					Lot Clean & Rough Grade	

STAGE 7	STAGE 8	STAGE 9	STAGE 10	STAGE 11	STAGE 12	TOTAL
Drywall Stage	**Int. Trim & Cabinets**	**Final Grade**	**MEP Trim**	**Instection Ready**	**Buyer Walks**	**Points**
6	**7**	**10**	**9**	**7**	**4**	**85**
Poly-Seal I	Counter Flashing	Blown Attic Insultation	Final Grade Survey	Window Inspection	Management Walk	
Wall Insulation	Gas/Electric Meters	Exterior Top Coat	Fences and Gates	Install Appliances	Final Inspections – All	Stage 0–6
Drywall	Interior Trim I	Countertops	Landscaping & Irrigation	Finished Floors	Buyer Demonstration	42 Points
Install Exterior Cladding	Poly-Seal II	Tile – Floor & Splash	Plumbing Trim	Garage Door Opener	Buyer Final Walk	
Garage Door Install	Set Cabinets	Interior Trim II	HVAC Trim & Start Up	Paint Final		Stages 7–12
Lot Clean & Rough Grade	Wall-Out & Enamel	Reglaze Windows	Electrical Trim	Exterior Clean Final		43 Points
	Bath Tile	Gutters	Low Voltage Trim	Interior Clean Final		
		Pour Flatwork	Weather Strip Ex. Drs.			
		Final Grade	Mirrors & Glass Doors			Apr-22
		Tree Installation				

THE STARTS PROCESS

STAGE 0

The starts process begins with the following three distinct tasks that must be completed:

1. **Buyer/Spec house selections:**
   » Finalizing buyer or spec home selections and options allows the purchasing team to create the job-specific budgets and plan set. The salesperson and design center perform most of this work, collaborating with the buyer.
2. **Lot-specific budget revisions:**
   » The builder has many responsibilities when creating the design of the building envelope and building pad for the placement of the home on the lot. The builder's design is memorialized in the job's final budget as part of the starts package.
3. **Permitting and approvals:**
   » With the selections and site plan finalized, the job is ready for submission to the city and all other required reviews and approvals.

4. **The End Goal**

    » The end goal is to deliver a complete and accurate job starts package to the builder, who then has everything needed to build the home to the plans and specs.

Decision-Making

- Buyers complete their color selections and plan options and make the required money deposits.
- Builder evaluates the site for conditions that may interfere with the planned improvements, such as manholes, utility boxes, fire hydrants, and other obstructions that may result in flipping the plan or making other site revisions.
- Builder evaluates the on-site conditions, designs the home's slab and pad, and estimates the lot's unique costs and slab budget changes.
- The builder submits lot-unique or site-specific budgets to the purchasing department and retains them for future use.
- Preliminary start notification (PSN) is released to the field one or two weeks prior to the lot being purchased, indicating to construction and sales that the home start is imminent.

Job-Specific Revisions

- **Architectural construction drawings** are revised to reflect all buyer options, plan changes, and/or changes to the job-specific plan set as needed to accommodate lot conditions.
- **Final site-specific engineering plans** (e.g., manual J, plot plan, foundation plan, truss layout) are revised to reflect all

buyer-selected plan options and all changes per the builder's submission for lot conditions.

- **Material takeoffs and labor budgets** are adjusted to reflect all buyer options/selections and all lot and/or slab budget changes per the builder's submission.

Approvals and Permits

- **Confirm that the planned home's impervious lot coverage** (concrete to dirt ratio) does not exceed the project's, city's, and/or county's applicable percentage coverage limits.
- Submit final drawings as needed from **the homeowners' association** and the developer's **architectural review committee** approvals, if any.
- Apply and receive **city or other governmental "permits"** per jurisdiction of the site.

Starts Package to the Field

- **Complete the lot purchase**—the "lot takedown"—if needed.
- **Issue the final starts notification** (FSN) to sales and construction.
- **Deliver the starts package** to the builder.

STAGE 0: THE STARTS PROCESS

PRESTART FIELD TASKS

Timing: The builder receives the Preliminary Starts Notification, which indicates that the start is imminent one or two weeks in advance of receiving the completed starts package and Final Start Notification (FSN).

Note: *Important:* No clearing or construction should commence until the FSN has been received and the lot is purchased.

Tasks: Upon receipt of the PSN, the builder performs the following tasks:

1. Review all records for this start (i.e., plan number and elevation, plan swing, legal description, and addresses) to confirm the information in the starts notification matches all other records in your system and on the building permit and all other information shown for this job.
2. Confirm that the plot plan as designed is compatible with existing fire hydrants, streetlights, storm sewer inlets, manhole covers, newly recorded easements, and anything that might interfere with driveway or house footprint and require a flip of the house swing.
3. Review the existing condition of the curbs, streets, and any existing sidewalks. Look for preexisting damage and document with photos to file.
4. Locate corner pins and existing electrical and cable lines on the homesite; contact the surveyor if needed.
5. Ribbon off (protect) trees and vegetation to be saved using four-foot-tall orange web fence as needed to protect tree(s) or per community requirements.
6. Set the T-pole and call for inspections. Schedule the water tap.
7. Confirm that the portable toilet, trash receptacles, and site signage are in place.

8. Confirm that the SWPPP (stormwater pollution prevention plan) erosion control was installed per plan and the concrete washout area is clearly marked and easily accessible by concrete trucks.
9. Design the home's finished floor (FF) elevation and slab design:
   - Coordinate soil borings for the structural engineer's design in communities in which soil tests are performed on every lot.
   - Calculate the best FF elevation for the home on this lot. Consider all options to achieve the necessary FF elevation at the necessary height to meet all flood or drainage and driveway slope requirements:
     - Slab must be 24 inches or higher than the curb.
     - Slab must be 24 inches or higher over the 500-year flood plain.
     - Slab must be at or above the engineer's set minimum.
10. Evaluate the driveway slope given the existing topography and calculate the garage elevation and the resulting driveway to slope the maximum allowed of 10 percent, or a 2-foot rise over 20 feet and a 2-foot-6-inch rise over 25 feet. Drop the garage as required for the driveway slope to be 10 percent or less.
11. If topography requires the garage to be dropped more than 48 inches to allow automobiles to enter and exit without scraping the bottom, notify the foundation engineer for job-specific drawings.
12. Design the construction plan of action for cutting or elevating the pad site to achieve the needed home and garage FF elevations.
13. Calculate the compactable fill/cut required as well as the compaction and testing, tree removal, and other lotunique grading costs, if any, to build the pad as designed on this unique lot.
14. Write all lot-unique budgets, documenting all quantities, calculations, notes, drawings, and documentation of the logic and resulting budgets using the company forms. Submit line-item budget revisions to the purchasing department for inclusion in the starts package budgets. Retain your records for reference later during construction.

STAGE 0: THE STARTS PROCESS

T-POLE

Timing: Upon receipt of the PSN

Task: The electrical contractor provides all materials and labor to install the temporary electrical pole for construction.

Note: The electrical contractor is responsible for the inspection and maintenance of the T-pole during construction.

Confirm the following:

1. The T-pole is set firmly in the ground to last for months and displays the address so that it can be easily read from the street.
2. A GFCI (ground fault circuit interrupter) for all 120-volt circuits was installed and grounded using a #6 copper wire.
3. On the home side of the meter, no aluminum wiring is less than #2.

STAGE 0: THE STARTS PROCESS

TREE PRESERVATION

Timing: Upon receipt of the PSN

Tasks:

- Surround saved trees using a 4-foot-high web fence around the drip line of the tree (matching limbs of the tree) to remain in place until the final grade stage.
- Construct protective wooden barriers as needed in tight situations; root pruning may be used based on tree size and distance from slab.

Notes: Native tree preservation is important to our communities and homeowners. We cannot guarantee the survival of native trees on the lot after construction, but we can do the following:

- Avoid compaction of soils within the tree drip line.
- Avoid excessive or dramatic grading that may alter existing drainage.
- Protect trees from chemical contamination:
    - » No fly ash, lime, lacquer thinner, paint, or gas near tree root zones
    - » No parking under trees

Confirm the following:

1. Trees designated to be saved are more than 10 feet from the planned house slab and are of a size or long-life species that warrant the effort to save them.
2. Understory within the vegetation ribbon remains undisturbed during the home's construction. Understory is defined as trees of 3-inch caliper or smaller as measured 12 inches above grade.
3. Tree limbs are pruned back to not closer than 18 inches from the future home. Paint with sealant all trimmed limbs and/or tree bark damaged during construction.

LOT PREPARATION STAGE

STAGE 1

- Receipt of the final starts notification (FSN)
    - » Review of the starts package
- Tree protection
- Pad preparation and construction (per builder's plan)
    - » Vegetation removal
    - » Tree removal and refill and compaction
    - » Building of the pad; excavation; filling and testing
    - » Pad elevation and construction certification
        - › By certified code inspector or surveyor
- Cutting the curb
- Installment of retaining walls (only as last resort)

The lot preparation stage involves clearing vegetation from the building envelope, removing, or protecting native trees as required, and installing the building pad as designed during the starts process to meet all elevational and drainage considerations in good condition to receive the house foundation. The pad is tested as required to certify Proctor density and elevational requirements.

STAGE 1: LOT PREPARATION STAGE

RECEIPT OF THE FINAL STARTS NOTIFICATION

Timing: Upon receipt of the FSN and starts package, which indicate to the builder that the lot is purchased and that the job is ready to start

Note: The starts package includes:

- Plot plan
- Architectural plans
- Engineered house and windstorm plans and foundation plans (including wet-set tiedowns for the foundation design)
- Building permit(s)
- Job-specific budgets
- Purchasing and buyer selections summary report
- Buyer plan change summary

Tasks:

1. Reconfirm calculated elevations for the pad/slab and garage elevation and review the plan by the certified code inspector.
2. Release the contractor to install the house pad per the approved design.
3. Verify selections and confirm current brick and/or stone and any other component materials with long lead times; release accordingly.

STAGE 1: LOT PREPARATION STAGE

TREE REMOVAL

Timing: Upon receipt of the FSN and starts package, the builder may clear and remove native trees.

Tasks:

- Trees designated to be removed are taken down either by earth-moving equipment or by another means in tight situations. In all cases, the stumps are removed from the ground. Tree trunks, limbs, and stumps are removed from the job site.
- The contractor refills these stump removal areas in 8-inch lifts and tests each lift for density. The builder reviews the records for each lift and retains them in the job's permanent file.

Notes:

- Existing trees and their stumps within 10 feet of the home must be removed and refilled in 8-inch lifts, compacted, and tested to 98 percent Proctor density (PD). Retain all tests records to the house file.
- Contractor imports or exports soil as needed and compacts to 98 percent PD to build the foundation pad at the builder's designed elevation in ready condition to receive the slab.
- Native tree preservation/protection is extremely important (see stage 0); see the following notes:
  - » Avoid compaction of soils within the tree drip line.
  - » Avoid excessive or dramatic grading that may alter existing drainage.
  - » Protect trees from chemical contamination:
    - › No fly ash, lime, lacquer thinner, paint, or gas near tree root zones
    - › No parking under trees

Confirm the following:

1. No tree remains within 10 feet of the building envelope (without engineering letter) to meet the buyer's request.
2. All stumps, limbs, and trunks are completely removed from the site.
3. After compaction, the previously wooded area is ready to construct the house's building pad.

STAGE 1: LOT PREPARATION STAGE

BUILDING PAD PREPARATION AND CONSTRUCTION

Timing: Upon receipt of the FSN and start package, which indicate that the lot is purchased, and the job is ready to start

Notes:

- The objective now is to clear the lot and install the building pad for the foundation according to the unique lot plan and costs estimate prepared by the builder during the starts process. Use those starts planning notes with your tractor contractor in laying out the project.
- If trees are near power lines, involve the power company first!
- An independent code inspector or surveyor will certify the house pad condition, elevation, and grade of the constructed pad.

Tasks:

- Tractor contractor shall remove all vegetation from the pad area and remove native trees per the builder's starts plan.
- All excess vegetation and excess soils should be removed from the site completely unless the soils are needed at a later stage.

Confirm the following:

1. All vegetation from the pad area was removed, and a 10-foot vegetation strip was left for erosion control along the curb.
2. Native soil or compactable fill material was placed in 8-inch lifts and compacted and tested in 8-inch lifts, with each achieving 95 percent Proctor density.
3. All native trees per the builder's plan were removed, filled, compacted, and tested and all reports were reviewed and saved to the files.

4. The constructed pad matches the design submitted during the starts process and will achieve a minimum of more than 8 inches of exposed foundation at the control point.
5. The overall lot is graded to the following guidelines:
    a. General grading standards:
        I. 8 to 10 inches of exposed slab
        II. 5 percent slope from the home to the main swales (6 inches in 10 feet)
        III. 2 percent minimum main swale slope to the street (24 inches in 100 feet)
6. House pad is level to plus/minus 1 inch across the entire pad.
7. Driveway slope and garage drop are per the builder's plan.
8. Zero excess dirt remains on site upon completion of the grading except as needed for retaining walls or other purposes planned at a later stage.
9. There was no damage to the curb, meter can, clean-outs, adjacent properties, T-pol (temporary electric pole), etc., at the completion of the tractor work.
10. If retaining walls are required on this homesite, the tractor contractor will cut a level surface for the walls' footings at this time.
11. The street is clean of any mud or debris.
12. The site is clean of any debris.
13. The independent code inspector has certified that the house pad condition and elevation are built to the correct elevation/grade per builder's plan and conform to governing minimums.

STAGE 1: LOT PREPARATION STAGE

CURB CUT

Timing: The curb cut ideally happens early to provide a site access point and have the cut curb debris ready for removal during the lot preparation stage of construction.

Task: The contractor machine-cuts the curb at the builder's designated location in a clean and appearance-acceptable manner.

Note: The curb cut contractor cuts one or more curbs per trip and leaves the cut curb on site for the builder to remove with other debris.

Confirm the following:

1. Prior to cutting, review the driveway location and swing with the contractor.
2. After completion, confirm the cosmetic finish of the cut. Expand the width of the cut as needed to remove any problem cuts.

STAGE 1: LOT PREPARATION STAGE

RETAINING WALLS

Timing: At the time of installing the house pad, the builder may elect to install retaining walls to achieve the best overall site plan.

Tasks: Contractor installs wall footings, builds the wall itself, and, behind the wall, installs drains as needed and backfills using native material.

Notes:

- Building a retaining wall is only one method of dealing with topography. Retaining walls affect lots on both sides and the drainage downhill from the lot; therefore, dropped beams and masonry ledges are the preferred method of addressing foundation exposure and topography issues. Consider all alternatives carefully when choosing retaining walls.
- Community approvals may be needed in advance to build retraining walls, and the builder should follow community-specific guidelines.
- Confirm all wall details, such as height and wall type, in advance with the buyer.
- Retaining wall is built on the low side of the property line if at all possible.
- Retaining walls lower than 4 feet (48 inches) high may be built without engineering sealed drawings. Walls must be designed by an engineer if taller than 48 feet or in places with built-in high-plasticity-index soils.

Confirm the following:

1. **All** manufacturer's guidelines and installation instructions were used when installing a retaining wall system.
2. Dry stack walls using at minimum 8-inch-thick stones/blocks have each course stepped back 1 inch from the course below.

3. Soil filter fabric is installed on the back side of the retaining wall with ABS-type plastic drainpipe covered with cloth and 0.75 inches of gravel and discharge remaining visible.
4. Weep holes are installed every 4 feet along the bottom of the wall when stones are in mortar.
5. Properly compact all backfill placed behind the wall, preferably with native material, or select fill in 2-inch-to-3-inch lifts, compacting each lift individually.
6. The homesite is clean and site signage is back in place when the wall is complete.

SLAB STAGE

- Forms
    - » Set forms and form survey.
    - » Form fill dirt and knockdown.
- Plumbing (electrical) grounds and sewer tap
- Plumbing grounds inspection
- Foundation makeup
    - » Beams and poly
    - » Electric and mechanical grounds (in-slab electrical/venting)
    - » Cables and static reinforcements installed
    - » Float forms
- Pre-pour inspections
    - » Plumbing and electrical inspection
    - » Independent code inspector's inspection, windstorm inspection (where applicable), etc.
    - » Engineer of record performs pre-pour inspection
    - » City pre-pour inspection

- Concrete placement
- Strip and dress foundation

The slab stage goal is to complete the house foundation per the engineer's design and per the slab location and elevation designed during the starts phase and inspected to conform to the city's requirements builder's design.

STAGE 2: SLAB STAGE

SET FORMS

Timing: When the house pad is constructed per the builder's plan and the lot is graded to achieve positive drainage away from the pad and off the lot and the pins are located and marked

Tasks:

- Using the identified homesite's corner pins, locate the building corners and lay out the entire building envelope.
- Construct the foundation form height to conform to specified finished floor (FF) elevation set by the builder's plan designed during the starts process.
- Because the forms are left open at the garage for access, prior to the cable installation, the contractor returns to close the forms left open at the garage.

Notes:

- Form height must conform to the builder's plan to achieve the minimum FF elevation per the plot plan and as high as required to allow more than a minimum of 8 inches of slab exposure at the control point.
- Early in the form-setting process, reconfirm that the driveway slope and transitions from the street to the drive and from the drive to the garage are manageable by the buyer's vehicle.

Confirm the following:

1. At the onset of setting the forms, the builder may confirm form FF elevation to be used and the house plan number, swing, elevation, wet set tiedowns drawings, and all plan options are reflected correctly on the contractor's field plan sets.
2. Form height, as measured from the top of the forms at the control point, conforms to the slab FF elevation set by the builder during the starts process.

3. Forms are level within 0.5 inches from the highest point to the lowest point anywhere on the forms.

Kicker spacing:

- Forms 18 inches tall or less are installed every 30 inches and at every joint.
- Forms taller than 18 inches are installed every 20 inches and at every joint.

4. Forms are square, straight, and plumb using the 6-8-10 rule of right triangle and a string line and laser level.
5. The driveway slope conforms to builder's plan to accommodate buyer's car(s).
6. If forms exceed 6 feet high, weep holes are installed along bottom.
7. The fireplaces projecting off masonry walls are dropped 1.5 inches from FF.
8. The forms remain open initially at the garage (preferably) for fill material and plumber access.
9. The form bottom edge is filled in as needed to contain the pour concrete.
10. Front porches are recessed per foundation plan, and, if brick pavers are planned, porches are dropped to accommodate the planned surface thickness.
11. Waste is scraped to the designated location.
12. All signage is replaced and clean.
13. Schedule the contractor to return to close the forms before the posttension cables are installed.

STAGE 2: SLAB STAGE

FORM SURVEY

Timing: As soon as the form setter starts work, the surveyor should be scheduled to survey the forms boards in place.

Task: The surveyor surveys and confirms that the slab location is per the plot plan and not encroaching on any easement or setback on any side.

Confirm that the form survey was performed without finding any encroachments and a copy is retained in the job's files.

STAGE 2: SLAB STAGE

PLUMBING GROUNDS AND SEWER TAP

Timing: When the forms are filled and level and before the beams are trenched, the plumbing contractor installs the plumbing grounds.

Tasks: Plumber trenches and installs all plumbing supply lines and sewer lines, connects each to the lines in the utility easements, and installs all plumbing boxes for drains at tubs and showers.

Notes:

- Plumber supplies all labor and materials and schedules and attends all city plumbing inspections. The plumber may provide a conduit for the electrician, or the electrician may install their own. In some municipalities, the electrician requires a separate inspection. Digging is done by hand or trencher as no backhoes are allowed inside forms.
- Prior to the starting the grounds, discuss the layout of the water distribution lines. Consider running supply lines not in the slab but rather up an exterior wall and through the floor joists.

Confirm the following:

1. Plumbing stub-ups match the floor plan of the house and are installed more than 12 inches above the FF, plumb, and with covered tops and secured with metal stakes.
2. The proper sewer slope and grade to the tap connection at the sewer line, and the green tag (passing inspection sticker) was obtained before covering.
3. Any pipe located in concrete is protected with plastic sleeves.
4. The main water line leading to the house is buried at least 18 inches deep. Contractor backfills and compacts sewer, water, and gas lines after inspections are passed.
5. Plumber has provided one temporary hose bib for construction on the outside wall.
6. Water pressure is maintained on water supply lines from this point forward.
7. Water cut-off was installed with valve and meter box secured at finished grade.

STAGE 2: SLAB STAGE

PLUMBING GROUNDS CITY INSPECTION

Timing: After the plumbing grounds are complete

Tasks: Plumber passes all municipal and other required plumbing inspections.

Notes: Either the plumber or the builder will call for and attend the city plumbing grounds inspection. In some municipalities, a separate electrical inspection is required on the conduit to the island.

Confirm the following:

1. Plumbing lines are pressurized and will remain so from this point forward.
2. The address is clearly visible from the street, often on the forms.
3. The sewer was tied in with a test ball or capped as required and is properly supported and bedded in sand entirely.
4. The location of the water heater to the farthest hot line fixture is less than 75 feet.
5. The location of the water softener loop and drain conforms to standard practice.
6. The water supply lines under the foundation are continuous, with no joints.
7. Water supply lines are capped approximately 12 inches above the finished floor and holding 100 pounds per square inch (psi) and with no kinks or leaks. Reconfirm after 24 hours.
8. No pipes were run parallel and inside a foundation beam.
9. Verify property lines are marked; ensure that there are string lines and a form survey available.
10. After passing inspections, the plumbing contractor backfills and compacts all water, sewer, and gas line undergrounds leading to the house.

STAGE 2: SLAB STAGE

FOUNDATION MAKEUP: BEAMS AND POLY

Timing: When the plumbing inspection is complete, the home is ready for beams and makeup.

Tasks: Makeup is a two-step process:

- Fill/cut fill material inside the forms to achieve the precise elevation, dig the beams, and install the vapor barrier.
- Install the posttension cable tendons and static reinforcing per the engineered foundation drawings.

Notes:

- Beams must reach a minimum of 6 inches into undisturbed soil unless the building pad is engineered (built in 8-inch lifts, compacted, and tested to 95 percent Proctor density).
- Beams are bagged when topography requires raising the pad elevation rather than digging down.
- The makeup crew must inform the builder of any observed plumbing damage!
- The form setter returns to close the forms after the beams and poly are complete and before the cables are installed.

Confirm the following:

1. Foundation beams are in the locations and to the depth and width specified in the engineer's drawings and specifications.
2. Beams are straight and walls are in good condition with no form or beam damage and drainage is restored out of forms for drainage.
3. Grade sand was added as needed to achieve a precise 4-inch foundation pad.
4. The 6-millimeter poly vapor barrier was installed to completely cover the slab and sides and bottoms of the beams following the contour of the beams, with no air pockets below the poly. All holes or cuts were repaired or taped as needed.

STAGE 2: SLAB STAGE

FOUNDATION MAKEUP: CABLE AND STATIC REINFORCEMENT INSTALLATION

Timing: The cable installation is part of the foundation makeup and performed by the same crew.

Tasks:

- Install posttension cables and other specified static reinforcement. The cable installer drills 1-inch holes in the forms as indicated on the foundation plans; installs the cables, nailing dead ends to the forms; and installs pocket formers, cat heads, and plastic chairs.
- Install static reinforcement (rebar) per the engineer's design.

Notes:

- The makeup package consists of the cables, rebar, mesh (as specified), chairs, pocket formers, and cat heads.
- Cable tendons consist of 0.5-inch cables in plastic sheathing, individually tagged and numbered to correspond to the installation diagram. Tendons are anchored permanently at one end, called the *dead end*, and stressed from the other end, called the *live end*.
- The individually numbered cables are installed following the precise locations and counts depicted in the shop drawings, and other steel reinforcing is installed to match the engineer's drawings.

Confirm the following:

1. Installation of all cables by number is per the installation diagram and rebar and other static reinforcement was installed matching the engineer's plans.
2. All slab tendon anchors are placed 4 inches down from the surface of the slab.

3. Cables are installed at least 6 inches from exterior building corners and a minimum of 12 inches of exposed cable are draped outside the forms.
4. Chairs spaced a maximum of 3 feet (36 inches) apart and tied to tendons with wire at each intersection and wherever cables touch poly, forms, or pipes.
5. Clearance is 0.75 inches where dead-end anchors are secured to the form boards between the tendon tail inside the hole drilled into the form board.
6. S-hooks were used as needed in exterior beams to support suspended cables.
7. Cables are tied together at every intersection to avoid movement during concrete placement.
8. There is no exposed cable from damaged sheathing; repair the cable sheathing as needed to completely protect the cables.
9. Where draped beam cables cross, the cables were tied to avoid horizontal movement.
10. Inspect and confirm rebar and other static reinforcement is installed as drawn and specified.
11. Poly completely covers bottom of beams and follows the contour of the beams with no air pockets below the poly and all holes were repaired or taped as needed.

STAGE 2: SLAB STAGE

FOUNDATION MAKEUP: FLOAT FORMS

Timing: The forms may be floated only after the cable and static reinforcement installation is complete.

Tasks: The contractor installs float material to achieve all specified slab recesses, porch/patio drops, garages, man-way doors, and brick ledges, and any other designed foundation recesses or drops.

Notes:

- A surveyor's instrument should be used to reconfirm and ensure form's levelness after installation of the float material.
- Reconfirm the slab's squareness near the garage after closing the forms and installing the float.

Confirm the following:

1. Slab is level within the tolerance of 0.5 inches from the highest to the lowest point on the slab.
2. Locations and dimensions for brick ledges and other raised or dropped areas are per the floor plan and the builder's submitted prestart plan and budgets.
3. All floats were properly secured to ensure levelness and squareness.
4. Garage car door floats allow for garage door railing and the garage door will be able to run to the bottom of apron float, flush with the slab.
5. Cables near float areas have good clearance between float material and the cables; be sure they are never touching.
6. The entry and back porch have a positive slope away from the home and door.
7. The garage has a minimum slope, or drop, of 2 inches from the back to the front of the garage.

STAGE 2: SLAB STAGE

PRE-POUR INSPECTIONS: MECHANICAL AND ELECTRICAL INSPECTION

Timing: After the cable and float forms have been installed and before having the foundation is inspected, the electrical and mechanical underground are installed and inspected.

Tasks: Installation and city inspection of all applicable electrical or HVAC (heating, ventilation, and air conditioning) elements running inside the slab

Notes:

- Check the plans for kitchen islands, floor plugs, underground dryer vents or downdraft cooktops, or other venting that may be required through the slab.
- Any 6-inch PVC venting pipes installed requires the doubling of the beam size.

Confirm the following:

1. The proper mechanical locations within the slab.
2. Venting through the slab is placed to allow for concrete to surround vent.
3. Proper depth of concrete (4 inches) where pipes are installed under poly.
4. All electrical conduit pipe joints are properly glued.
5. All conduit and venting are secured properly to withstand concrete placement.
6. All cuts or damage to poly is taped closed.

STAGE 2: SLAB STAGE

PRE-POUR FOUNDATION INSPECTIONS: ICI, WINDSTORM, ENGINEER, CITY

Timing: After the forms are floated and the makeup process is complete, the applicable inspectors should be notified that the house is ready for various foundation inspections.

Tasks: Schedule and pass the designing engineer of record's, the independent code inspector's (ICI), and the city's pre-pour inspections. Builder must obtain passing inspections prior to scheduling the pour date.

Notes:

- Reconfirm the form survey is complete and without issues before the foundation pour.
- Have the (revised) engineered plans and the building and plumbing permit(s) available at the time of the inspection.
- Meet all inspectors on site for the inspection. Be present and professional.
- Do not proceed until all inspections and re-inspections are passed!

Confirm the following:

1. The property lines are identified, the stakes are visible, and the form survey is available.
2. The poly follows contour of beams, all beam walls are straight, and there is **no** water inside the beams.
3. All cable intersections remain firmly tied.
4. All exterior door locations are clearly marked on the forms.
5. If the garage was dropped, the waterproofing/French drains are installed.
6. The address is prominently posted and visible from the street.
7. Confirm all trash is in designated receptacles for a debris-free job site.

STAGE 2: SLAB STAGE

CONCRETE PLACEMENT

Timing: Only after all inspections are complete and passed

Task: Place and finish the concrete to achieve the designed slab.

Notes:

- **Concrete is the single most expensive component in the construction of a home.** Accordingly, it is imperative that the concrete placement be properly planned and executed.
- **The weather should be clear of rain** (40 percent chance of rain or less and falling) and at least 40 degrees and rising at the start of the pour. On a 100-degree day, a cold joint can form in 30 to 50 minutes. On hot days, confirm that you have adequate labor to manage one concrete truck every eight to 10 minutes.
- **Place concrete as early in the day as possible.** Arrange to have the contractor on site an hour before the pour. Never allow concrete to be delivered if the contractor is not waiting on site.
- **The "wet installation" locations of tie-down and straps** are often shown on the foundation details, not the foundation plans themselves. Have all wet sets on site on the day of the pour.
- **The foundation design engineer will perform an inspection** at this point and attend the concrete placement. The engineer's report from the pre-pour and pour inspection must be retained in the job's permanent files.
- **Avoiding surface cracking:** Shrinkage cracking results when freshly placed slabs are subjected to a rapid loss of moisture caused by wind, low humidity, and/or high temperature. Shrinkage cracks are rarely an issue for the integrity of the slab but are unsightly and cause concerns for customers.
- **To help minimize shrinkage cracks:** 1. Dampen the subgrade and form boards; 2. use a concrete curing agent; and 3. cover concrete with wet burlap/poly, keeping the slab wet and covered for several days.

Tools for Builder's Final Pre-Pour Inspection:

- All architectural and foundation engineering drawings, including p ot plan
- Two tape measures: 25 feet and 100 feet
- Level 4+ inches
- Marker pen and spray can

Confirm the following:

1. The required maximum slump, generally indicated on the foundation plan at 5 to 6 inches. No water is to be added to the concrete mix without the builder's direct approval. Never add water beyond a 5-to 6-inch slump!
2. Cables and pipes must not be displaced during the pour.
3. No concrete has entered any plumbing stub-outs.
4. Form boards are thoroughly vibrated during the pour to avoid honeycombing of the slab.
5. The concrete is to be leveled by means of a screed board and bull floats.
6. Slab surface was completely machine-troweled to the required finish, with the garage and living areas the same.
7. All edges of garage beams, entries, and thresholds were finished with an edger and the drop was sponge-floated to a straight and smooth finish.
8. Slab surface is not uneven by more than 3/8 inches in any 32-inch radius (i.e., dip or hump in the center); there should not be more than a 0.5-inch deviation in level over any 50-inch span.
9. Confirm placement of all anchor bolts, windstorm straps, dogeared tiedowns, and other wet-set anchor locations.
10. No bolts closer than 12-inches on either side of a door. Bolts are placed in 3-foot on-center (OC) for two stories and 4-foot OC for single-story homes, unless otherwise specified by windstorm or other code requirements as noted in the engineering drawings.

11. Water meter and sewer inlet are not damaged. Replace stakes if needed.
12. Concrete washout is at designated location only. All overspills must be picked up.
13. Confirm the contractor is scheduled to wreck the forms the following day. The float forms are not pulled before the concrete has been set up adequately.
14. The cable company is contracted with the pour date to schedule the cable stressing.

STAGE 2: SLAB STAGE

STRIP AND DRESS FOUNDATION

Timing: The day after the foundation concrete is placed, the concrete contractor may strip the forms from the slab.

Tasks: Strip the forms from the slab, stack the material, and dress any slab imperfections (honeycombing).

Notes:

- No payment is made for this task as it is part of the slab contractor's overall scope of work.
- Forms are **pulled** from the slab and not broken away. Chipped slab edges are a concern and may have to be repaired in highly visible areas. The likely repair uses tile, slate, epoxy surfacing materials, or other similar products. This is an avoidable problem, and the individual doing this work should be observed at the onset of work and coached by the builder as needed.

Confirm the following:

1. All form boards and stakes were removed.
2. Forms were removed in a manner to avoid chipping the slab surface.
3. All honeycombs were grouted immediately, with **no** steel exposed.
4. All imperfections and honeycombing in brick ledges and surfaces are floated.
5. All components and trash were scrapped to the designated location or receptacle.
6. Usable material is stacked in the garage area.
7. All SWPPP (stormwater pollution prevention plan) erosion control is installed and the signage is clean, in good condition, and replaced at the street.

ROUGH GRADE STAGE

- Release frame lumber package (per current lead times)
  - » Main walls
  - » Beams and I joists
  - » Subfloor
  - » Rafters
- Stress foundation cables (more than seven days after pour)
- Clean and rough grade the lot

The rough grade stage is centered on the establishing positive drainage around the new slab to conform to the drainage pattern shown on the plot plan or as designed by the builder during the starts phase. Establishing positive drainage around the home at this time, prior to any further work commencing on site, is important for the safety and efficiency of the framing and other crews that follow. The rough grade must be established prior to the framing lumber's arrival.

STAGE 3: ROUGH GRADE STAGE

STRESS FOUNDATION CABLES

Timing: The cable supplier is notified on the date of the pour and thereby schedules the stressing 7 to 14 days after concrete placement. Stressing may occur beyond the fourteenth day, but the extending cables will interfere with the work of other contractors.

Tasks: Cable supplier stresses the cables to the specified tension; removes the catheads, pocket formers, and protruding nails; and grouts both live and dead ends flush with the slab.

Note: Cable supplier documents the stressing (stressing certificate), and this becomes part of the home's permanent records.

Confirm the following:

1. All cable holes are grouted flush after cables are stressed.
2. Protruding nails from castings on both live and dead ends are cut flush.
3. Pocket formers, cable ends, and all other remaining trash from the stressing are placed in the provided trash receptacle.
4. The stressing certificate is retained in the home's permanent file.

STAGE 3: ROUGH GRADE STAGE

ROUGH CLEAN AND GRADE

Timing: After the foundation forms are wrecked and prior to the frame material being delivered, the house is ready for the rough grade.

Tasks:

- The tractor contractor reestablishes the drainage pattern away from the slab and off the homesite by cutting drainage swales per the plot plan or site drainage plan to ensure drainage off the lot.
- Rough grading is required to provide safe and partially level access to the home for future contractors.

Notes:

- Never deliver framing material before the rough grade is complete.
- Builder provides the fill material used at the rough grade by the tractor contractor, who shall follow these grading standards:
  - » From slab to swale: 5 percent minimum (falls 6 inches in 10 feet)
  - » In the swales: 2 percent minimum (falls more than 24 inches in 100 feet)

Confirm the following:

1. Grading allows approximately 8 to 10 inches of exposed slab.
   - » Account for final grade soil and sod added at later stages.
2. Grading falls away from slab, dropping a minimum of 6 inches to the property line or to 10 feet off slab.
   - » Measure 6-inch elevation drop to property line relative to the grade of the ground against the poured slab.
3. Confirm all swales cut into the lot meet at smooth transitions and conform to the drainage plan illustrated on the plot plan or builder's prestart plan.

4. Confirm sufficient fill material is delivered for spreading around the lot to allow soil for the sidewalk that will be cut into at a future stage.
5. Confirm the finished lot is smooth from the slab to the curb. There is hand grading at all transitions between swales and around the perimeter of the slab, meter boxes, manholes, electrical transformers, or other features of the lot.
6. There is **no** standing water (bird baths) on the lot!
7. Confirm SWPPP (stormwater pollution prevention plan) erosion controls and that signage is undamaged, clean, and in place and there is no dirt in the street.

FRAME STAGE

STAGE
4

- Release next-stage materials per current lead times:
  - Windows and exterior doors
  - Cornice and exterior sheathing
- Frame the house:
  - Chalk out the slab and review.
  - Frame the house.

The frame stage is the construction of the home's two-by structural skeleton. This process is part science and part art. The builder plays a key role early in the process by examining the chalk lines before framing starts and upon completion and inspection of the framer's work.

STAGE 4: FRAME STAGE

FRAME

Timing: After the slab is poured and the rough grade is complete, it is time to frame the house.

Tasks:

- After reviewing the plan and selected options as shown on the plan, the framing contractor chalks out the floor plan onto the foundation. Builder and the framing contractor evaluate and agree on all aspects and solutions of the floor plan layout as chalked out on the existing slab.
- The frame contractor builds the walls, floors, and roof frame per the architectural construction drawings and the engineered framing, strapping, and bracing plan set provided, per International Residential Codes (IRC) and as agreed upon with the builder.

Notes:

- Prior to chalking out the slab, builder and contractor review the floor plan and selected plan options and confirm the heating, ventilation, and air conditioning (HVAC) equipment platform and air conditioning (AC) plenum and fire blocking locations and review known tight AC duct locations that will need special attention. Work with your framing contractor early to agree on all framing issues and solutions.
- If the slab is not perfectly square, the sole plate needs to be placed inside or extending (only slightly) outside the foundation as needed to ensure that every room is framed square.
- Wall and plumbing layout are reviewed at chalk-out for fit inside the walls. Design the out-of-wall solution (if needed) at the time of chalk-out, by moving either the wall or the plumbing.

- Correct framing will ensure smooth and predictable completion of the home. Plan to visit the home several times daily during framing.

Confirm the following:

1. Styrofoam sole plate sealer (either 3.5 or 5.5 inches, depending on stud size) is installed under all exterior and garage wall plate locations.
2. Wall framing is plumb and square to a standard tolerance of 0.25 inches over 32 inches maximum allowed as measured by a dropped plumb line.
3. Every stud is secured with two 16-penny nails at the top and three nails at the bottom plate to prevent the stud from twisting at the bottom.
4. All beams and trusses are supported to the slab by studs or posts that are the total width of the beams or trusses being supported. Confirm double, triple, etc., studs as required under all beams, valleys, ridges, and bracing loads.
5. Where on-center specifications allow, all framing members will be stacked, with joists over studs and rafters face to face with the joists. All loose joist ends are secured with a strongback run to adjacent, secure joists.
6. No stud bows beyond 3/16 inches in a 32-inch drop as measured horizontally by plumb line. Contractor may repair bowed studs by cutting into the bowed stud up to 25 percent of the stud's width and then apply a nail-and-block technique. Bowed stud repairs are not necessary when using finger-jointed studs.
7. Confirm all window header heights match the architectural drawings, especially for arch top, transom, or other specialty windows.
8. No sill height in any bedroom is higher than 42 inches, per IRC for egress.
9. Framed rough openings for all tubs and precast shower pans are at or slightly greater than the minimum size per the plan.
10. Front door rough openings with sidelights are framed to allow the sidelight(s) header height to be flush with the top of the brick mould, allowing for a single lintel across both the door and window(s).
11. Corner wind bracing was installed to comply with the appropriate codes, generally required for all exterior outside and inside corners: OSB (oriented strand

board) extending 4 feet or more from each corner. Where windows or doors do not allow bracing, structural sheathing is the common solution.

12. In windstorm zones 1 and 2 (or in municipalities requiring the same), the framer installed the complete continuous strap system from the plate onto the studs and onto the top double plate and using connectors and straps through to the rafters and ridge. The framer installed the complete system per the engineered plans and in accordance with windstorm zones and codes.
13. All joists are crowned up, with the bowing side to the sky.
14. Framing allows for all recess lighting and directional can lights to be installed per the electrical plan.
15. Joist hangers are installed on all joists greater than 4 feet in length and nailed 100 percent in every hole provided on the hanger.
16. The joist lap over walls does not exceed 2 inches.
17. Second floor subflooring was installed with staggered joints and glued, nailed, and screwed according to company specifications.
18. Fire blocking was installed as required; for example, all joist cavities adjoining unconditioned space are fire-blocked to prevent air infiltration into conditioned spaces.
19. If two studded corners are used in framing, then drywall clips or deadwood were installed as required for nail blocking for drywall installation.
20. Strong backs were installed atop of the joists for every 8 feet of joist span and are fully secured to other adjacent framing members.
21. HVAC platform and walkway were built per local code, minimally a 30-inch platform with a 24-inch walkway not to exceed 20 feet in length. Confirm the location and its structural integrity of the joists supporting the platforms.
22. Water heater location: If it is in the garage, the platform was built per standard; if it is in the attic, it is accessible and located directly over a load-bearing wall and raised to accommodate 18 inches of insulation below.
23. Attic stair opening is double-framed on all four sides and with joist hangers if required.

24. Insulation dams (vertical barriers) were built or installed in the attic to hold blown insulation in place at key locations at the specified depth.
25. Roof bracing bears down on locations as specified on the framing cut sheets and never over a single joist or door or window opening.
26. All rafters are crowned up, bowing upside to the sky.
27. The roof ridge is one size—2 inches—larger than the rafters and installed level and double-braced. Any ridge break is braced directly underneath.
28. Spliced ridges, rafters, hips, and valleys are fully braced using a V-splice or dovetail method, centering a 4-inch plywood gusset to extend 2 feet past on both sides of the splice. The plywood gusset is glued and nailed and braced directly below.
29. Purlins are the same dimension as the rafter, placed under every other rafter, and installed square to the rafter and never less than 45 degrees or more than 25 degrees to the horizontal.
30. The cricket behind the fireplace was installed at least 4 inches past chimney on both sides and pitched at 12:12.
31. The frame was swept clean, and all scrap was placed in the designated receptacle.
32. Unused material was stacked in the garage. Contact purchasing if take-off corrections are necessary.

MOST COMMON INSPECTION FAILURES INCLUDE:

- All joist hangers installed
- All beams bolted completely
- All safety straps installed
- Fire blocking complete
- Load-bearing members braced to the plate
- Top plates overlapping and tied correctly
- Home swept and clean and extra material stacked neatly in garage

CORNICE AND SHINGLES STAGE

MEASURE FOR CABINETS

- Measure for cabinets
- City and licensed code inspector's (LCI) frame inspections
- Termite protection spray
- Sheathing installation
- Cornice installation
- Roof deck installation
- Exterior paint prime
- Fireplace set
- Roof underlayment and shingles
- Cleaning and rough grade of the lot

The cornice and shingles stage first involves the inspection process for the new frame and for the measurement of the cabinets as needed. The cornice crew follows the sheathing-cornice-deck progression leading to the roof underlayment and shingles, these collective steps will weatherproof, or dry-in, the home. After which the painter can prime the exterior cornice and the fireplace may be installed. As with many stages, this stage ends with a lot clean and rough grade to reestablish positive drainage around the house and off the lot.

STAGE 5: CORNICE AND SHINGLES STAGE

MEASURE FOR CABINETS

Timing: Cabinet lead times vary from year to year, but it is never too early to measure for cabinets. In some years measuring for cabinets could wait till after drywall, stage 7. However, when the lead times are unpredictable there no reason to wait after the framing is complete.

Task: Cabinet contractor measures all critical captured dimensions and chalks out all the cabinets on the slab using the cabinet shop drawings or the plan's interior elevations.

Note: If the contractor's shop drawings do not substantially agree with the plan's interior elevation, contact the purchasing and/or architectural department for clarification.

Confirm that the builder and contractor have reviewed any identified issues and agree that the cabinets as chalked on the slab are correct.

STAGE 5: CORNICE AND SHINGLES STAGE

CITY AND/OR LICENSED CODE INSPECTION: FRAME

Timing: Schedule the LCI for the frame inspection as soon as the framing is complete.

Tasks: The LCI inspects the frame for code violations and confirms all beam sizes and locations to ensure compliance with both local code and International Residential Code (IRC) and conformity with the engineer's design and specifications.

Note: The LCI frame inspection report becomes part of the home permanent record.

Confirm the following:

1. All current plans and engineering drawings pertaining to the house are available to the LCI during the inspection.
2. Once the house has passed inspection, retain the passing tag and/or certificate in the permanent house files.
3. If framing changes need to be made, either perform the work and have the home reinspected or confirm the work is complete and send photos to the inspector. Keep all correspondence and photos on the identified defect and the completed solution implemented in the permanent house files.

STAGE 5: CORNICE AND SHINGLES STAGE

TERMITE PROTECTION SPRAY

Timing: When the 2-inch framing is complete and the roof deck is in place (before insulation), the termite treatment contractor may treat the frame.

Task: Contractor completely sprays the termite pretreatment mixture on all framing members approximately 24 inches high on the entire first floor frame.

Notes:

- The wood is treated using a mineral salt borate solution, which kills the termites if they consume the treated wood.
- The greenish treatment dye is for tracking purposes, ensuring that all members are completely covered. The same chemical solution without the color may be used as needed near and around windows and doors to avoid staining.

Confirm that all first-floor framing members are treated approximately 24 inches up the studs.

STAGE 5: CORNICE AND SHINGLES STAGE

SHEATHING INSTALLATION

Timing: When the frame is complete, the home is ready for sheathing.

Tasks: The installing contractor installs the flashing over the brick ledge and up 12 inches up and onto the wall and then secures the sheathing to the entire surface of the home and wraps the corners with poly or felt. If house wrap is used, it is installed at this time.

Note: The sheathing acts as a vertical drainage plain and moisture/air barrier.

Confirm the following:

1. Sheathing is attached to every stud using fasteners installed vertically or according to the manufacturer's specifications for that product.
2. Framing member (stud or blocking) is located behind every vertical and horizontal sheathing joint.
3. Poly flashing is applied 12 inches up the studs behind the sheathing and draped to cover the brick ledge completely.
4. Z-joint flashing is installed continuously at every horizontal sheathing joint when oriented strand board (OSB) sheathing is used. Some sheathing products are designed to overlap on top of the horizontal joints. All sheathing joints and building corners overlap from the top and are taped closed using the manufacturer's recommended tape and installed per the manufacturer's design.
5. Sheathing covers all headers and was installed to extend up to and over the top plate, with no gap.
6. Blackout sheathing was installed behind all blind windows and vents/louvers.
7. Window Z flashing was installed above each window, poly-wrapped, and taped on both sides and then taped along the top last.

STAGE 5: CORNICE AND SHINGLES STAGE

CORNICE INSTALLATION

Timing: The house is ready for cornice once the sheathing application is complete.

Tasks: Build and install the following: Install any siding products, build the box cornice (lookouts, fascia, soffit), and install exterior doors, ornamental vents/louvers, and all windows.

Notes:

- The contractor supplies the galvanized nails for the siding (no staples). Cornice contractor installs the appropriate metal flashing as required for each application.
- Much of the cornice materials are fiber cement products, which are extremely specific to the pressure permitted by the installers nail gun. Overdriven nails are unsightly and may void the warranty on the siding and other fiber cement products.

Confirm the following:

1. Mitered joints at both the brick frieze and facia are tight and even from top to bottom.
2. Perforated soffit is continuous (except in gables) and nailed at every lookout, and H spacers were installed at each joint; no drooping corners.
3. Brick frieze perfectly lines up vertically over the brick ledge.
4. Window trim is one solid piece at the top, extending fully over the side trim; sides run solid to the bottom of the bottom trim, and the bottom trim is centered and short.
5. Facia is level and straight to the eye; confirm with a level.
6. Metal lintels installed over windows at masonry locations are level and blocked and nailed or bolted into place.
7. Siding boards are nailed at every stud with proper pressure and recess.

8. Siding was installed straight, level, and smooth, with no visible bows.
9. Siding was installed to extend 1.5 inches below the bottom plate but never resting on concrete.
10. Window installation:
    a. Windows were left in the locked position while being installed.
    b. Windows are plumb and square and centered in the rough opening.
    c. Windows were installed right side up.
    d. Twelve-inch on-center (OC) nailing into the nailing flange.
    e. Window operation is smooth, and windows close tight and firm.
    f. Peal and seal window tape was installed wrinkle-free over a clean, dry surface to cover the two sides first and then the top of all windows.
11. Exterior doors are set and blocked with three sides plumb and square and tapped and sealed on the top and both sides.
12. Shutters, timbers, and columns or posts shown on elevation were installed per elevational plan.
13. There is proper installation and overlap of all J-type and L-type 6-inch-by-6-inch metal flashing at step gables, boxed chimney chase, and other locations as needed.
14. All signage was replaced, and all SWPPP (stormwater pollution prevention plan) erosion control was restored as needed.

STAGE 5: CORNICE AND SHINGLES STAGE

ROOF DECKING

Timing: As soon as the facia is installed, the decking may be installed.

Task: Decking contractor installs the builder's roof deck and H clips and provides the nails and all tools for this task.

Notes:

- Because decking material is a theft target, schedule delivery for the first morning run on the day needed and never delivered on Friday afternoon for Saturday.
- Inspect the home for overcrowded rafters and discuss the cut-and-shave solution with the decker. (Cut/shave requires an engineering letter to be nailed to the repair.)

Confirm the following:

1. The OSB decking's factory label and radiant barrier were installed with the brand label facing down, visible from inside the home, with the textured side up.
2. Full sheets should be used first, with cutoffs used only to complete areas.
3. The 8-foot-long side of the decking runs perpendicular to the rafter, and all vertical joints are staggered and meet on a rafter or are braced.
4. Half-inch aluminum H spacing clips were installed with two per side and at all joints and over rafters, ridges, or valley.
5. Decking is flush to the inside edge of the 1x facia and not overlapping.
6. Crickets were installed and decked and nailed secure behind the fireplace.
7. Nails, **not** staples, were used for installation.
8. Roofer installed underlayment over the decking immediately upon completion of the deck.

STAGE 5: CORNICE AND SHINGLES STAGE

EXTERIOR PAINT PRIME

Timing: After the cornice is complete (preferably before shingles), the home is ready for the painting contractor to apply the exterior paint prime.

Task: All finished surfaces of the home (cornice, siding, trim) are prepped and primed, including all pre-primed materials.

Notes:

- The paint contractor provides all materials and labor to prep and prime the home.
- The materials specification is 100 percent acrylic paint and elastomeric caulk.
- Painting contractor must comply with all Environmental Protection Agency regulations in discarding all paint and paint-related chemicals.
- Prime only when the outside temperature is 40 degrees and rising.

Confirm the following:

1. All surfaces receiving paint are clean at the onset of the process.
2. The entire home is properly prepped with approved putty and caulk, resulting in smooth and even surfaces for the paint.
3. Painter uses only the approved elastomeric caulk; no latex caulk is allowed.
4. All surfaces are primed, including surfaces labeled *pre-primed*. The prime paint is among the approved prime colors and installed more than 3 milliliters thick when wet over all finished exterior surfaces.
5. There are no paint runs or overspray on shingles or windows.
6. All debris was picked up and placed in the designated trash container.

STAGE 5: CORNICE AND SHINGLES STAGE

FIREPLACE SET

Timing: The fireplace may be set after the frame, decking, and cornice are complete through to the chimney chase.

Tasks: The process involves two trips:

1. Place the fire box and flue and the termination cap top.
2. Install the fireplace doors and start up the direct vent fireplace or install the gas logs and test operate the system.

Note: Fireplace contractor is a turnkey contractor who provides all the required materials and labor.

Confirm the following:

1. Fireplace was placed on-center (OC) of its opening.
2. Fireplace was set on 8inch blocks or on the ground to match the height shown on the home's interior elevational drawings.
3. Fireplace was fastened solid to the framing members at on-center intervals per manufacturer's specification.
4. Fireplace is offset at least 2 inches from any combustible framing materials.
5. Fire stop was installed along all ceiling lines.
6. Fresh air kit was installed and is visible from the ground and open (if applicable).
7. Chase cap was installed with the collar sealed.
8. Direct vent fireplace manufacturer's installation was followed exactly. Direct vent venting must be at least 24 inches from any hip or valley on the roof.
9. All trash was placed in the designated receptacle, and the home was swept.

STAGE 5: CORNICE AND SHINGLES STAGE

SHINGLES AND ROOF UNDERLAYMENT

Timing: The home is ready for underlayment and shingles after the decking is complete.

Tasks:

- The roofing contractor installs the approved underlayment immediately upon completion of the decking.
- The roofer installs the shingles, drip edge, flashings, and plumber-supplied vents.

Notes:

- All roofing workers must follow Occupational Safety and Health Administration (OSHA) regulations and tie-off using D rings provided to tie-off or other approved fall protection systems when working on the roof.
- The home is dried-in once the underlayment is installed.
- Roofer must notify builder of any flashing concerns where additional flashing or additional pans are recommended.

Confirm the following:

1. Shingle starter course overhangs the facia by 0.75 inches.
2. Shingle drip lines create a uniform pattern throughout the entire roof.
3. Installation follows manufacturer's shingle fastener specifications:
    a. Four 1.5-inch roofing nails per shingle with minimum 0.75-inch penetration.
    b. High-wind zones (zones 1 and 2) and per local municipality code require six nails. Always confirm local requirements.
    c. Nails placed in the nailing strip only.
    d. No exposed nails in the roof.
    e. Staples are never allowed.

4. Valley flashing is 28-gage and runs continuously the entire valley length.
5. Shingles are held 2 inches off any side wall and overlapping the installed J flashing.
6. Roof vents and jack flashing are nailed with galvanized nails and sealed with roofing cement/sealant over any exposed nails. Vent boots are nailed at the center of the bottom of the boot. Note that the pitch boot must match the roof pitch.
7. HVAC (heating, ventilation, and air conditioning) venting was installed and sealed through the roof deck.
8. Shingles and ridge shingles lie smooth and in place.
9. There are no nails outside the nailing strip.
10. Roof is clean, there is no visible mud or footprints, and all debris is in the proper containers.
11. Site signage is clean and restored along the curb, SWPPP (stormwater pollution prevention plan) erosion control is reinstalled, and no dirt is in the street.

STAGE 5: CORNICE AND SHINGLES STAGE

TRACTOR: ROUGH CLEAN AND GRADE AFTER SHINGLES

Timing: After completion of the cornice and shingles

Tasks:

- Contractor performs a general site cleaning of all micro-trash and any other waste of any kind prior to performing any tractor work. All cornice scraps, scrap shingles, felt, and packaging are removed from the site by the contractor.
- Usable building material identified at this time is stacked into the garage.
- Contractor rough grades the lot to reestablished positive drainage off the lot according to the plot plan or the builder's plan designed at the starts process.

Note: Do not allow trash to be simply graded into the lot. This trash shows up later during final grading or is found by the homeowner after closing.

Confirm the following:

1. Lot is graded smooth with no birdbaths and positive drainage off the lot.
2. Trash piles and trash containers are empty, clean, and back in place for future use.
3. There is no random or blown trash on adjacent properties.
4. Usable material is stacked neatly in the garage.
5. SWPPP erosion control is back in place, with no dirt left in the street.
6. Signage is clean and restored back to the street.

MEP ROUGH-IN STAGE

STAGE 6

- Plumbing and gas top-out
- Heating, ventilation, and air conditioning (HVAC) rough-in
- Electrical rough-in
- Electrical underground
- Low-voltage wiring rough-in
- Frame clean and vacuum
- Precover inspections, various
- Buyer precover walk (buyer's option)

The mechanical, electrical, and plumbing (MEP) rough-in is the phase during which the builder installs the walls with all specified systems following the logical progression of plumbing-HVAC-electrical. This critical stage is the last chance to install anything inside the wall and for various precover inspections by the buyer and city. Prior to any inspections, the home itself and all frame cavities are thoroughly cleaned. At the completion of this stage, the home is approved and ready for wall insulation and drywall.

STAGE 6: MEP ROUGH-IN STAGE

PLUMBING AND GAS TOP-OUT

Timing: The house is ready for plumbing top-out after the framing, cornice, and fireplace are complete and the shingles—or roof underlayment, at a minimum—are installed.

Tasks:

- The plumber connects the house to the water meter; installs all hot and cold-water PEX (polyethylene pipe) supply lines, tubs, shower pans, and stub-out for all plumbing fixtures and drains; and leaves roof jacks vents for the roofer.
- Gas lines are installed and pressurized to maintain a pressure test to 15 pounds per square in (psi), or the psi as required for that city's inspections.

Notes:

- The gas utility line underground is run and stubbed to within 12 inches of the slab by the gas utility company.
- No gas stub-out may be located less than 36 inches to any door, window, breaker panel, or HVAC condenser.
- Plumber must conform to the I-joist boring sizes allowed and specified by the I-joist manufacturer.

Confirm the following:

1. Maximum fixtures per line are as follows:
    a. For cold outlets: three fixtures per 0.5-inch line
    b. For hot outlets: two fixtures per line
2. There was no excessive cutting on floor joists, no top or bottom flanges are cut or damaged, and no holes were cut through beams.

3. Hose bibs were installed per plan and options and perpendicular to the wall at a reasonable distance from the ground. Bibs are nailed secure to the blocking inside the wall.
4. The whole house water cutoff valve was installed per plan, usually in the garage or utility room.
5. No water lines, sanitary sewer lines, or gas lines run through or across the attic catwalk or access openings or in fireplace chases.
6. Shower heads and second-floor commode flanges are completely secured with brass screws.
7. Valve control locations in all showers are conveniently accessible by the user. Usually the mixer valves are centered on tubs and most accessible at the entrance of the showers.
8. PEX water lines are secured with PEX fasteners or wood wedges when penetrating wood framing members to prevent pipe vibration noises.
9. Washing machine and ice maker rough-in boxes are level, blocked, and nailed secure on both sides.
10. No water supply lines run above the insulation line. The only exception is the line leading to the water heater, which is individually insulated completely to the water heater.
11. Water heater installation:
    a. In the attic, the water heater was installed on an approximately 2-feet-by-4-feet platform sitting on top of a load-bearing wall or a beam. Builder must confirm the structural integrity of the framing below an attic platform.
    b. In the garage, the water heater was also installed on a platform built by the framer, per the plan.
12. Proper-sized nail guards were installed throughout the home as needed to protect water lines. Note: Missing nail guards are a common missed item on inspections.

13. Bathtubs and shower pans are sitting level to the ground, not rocking when walked upon, and covered with protective liners.
14. Water supply lines are free of kinks and all turns made with the manufacture's PEX corner brace and have a smooth layout throughout all walls.
15. Water lines have maintained static pressure.
16. There are no overcut holes through the sheathing and nothing higher than 0.25 inches off any pipe. Repair sheathing as needed to reestablish the watertight envelope.
17. Water heaters were installed per the plan in the attic, or the garage or tankless water heaters were installed outside on the garage wall.
18. Water softener loop was installed per the plan, either in the garage next to the water heater or on an outside wall for outside installation.
19. Gas lines are installed and achieve and maintain pressure.
    - No gas lines are run outside the sheathing.
    - No concealed gas line unions are allowed.
20. Home was swept clean, and all debris was scrapped to the designated receptacle.

STAGE 6: MEP ROUGH-IN STAGE

HVAC ROUGH-IN

Timing: Start the HVAC rough-in once the plumbing top out is complete and the home is swept clean.

Tasks: The contractor furnishes all material and labor required to install and seal the ducts, attic mechanical units, plenums, and return airs as per the system's approved design.

Notes: Buyer options (e.g., additional windows or large vent hood, commercial appliances) may affect the performance of the system. Builder must alert the contractor to rerun a new manual J on the house if plan or specification changes significantly affect the system.

Design Specifications:

- The system is designed to cool the home to 75 degrees Fahrenheit. In extremely high temperature conditions, the system is designed to cool to a maximum of 20-degree temperature differential from the outside temperature as measured at 5 feet off the floor at the center of the room.
- Balance rooms to within a 3-degree temperature differential from the thermostat set temperature as measured at 5 feet off the floor and at the center of any room.

Confirm the following:

1. Air supplies were installed to each room, bath, closet, and hall per the approved design.
2. Return air or passive return air (jump ducts) is required in all rooms with an operable door. Rooms not requiring return air or jump ducts include utility rooms, bathrooms, and closets.

3. The primary HVAC drain connects to the plumbing drain. The secondary drain is run to and extending through the soffit and visible from an inside window but not on the front elevation of the home.
4. Flex ducts are run straight and horizontally as much as possible and supported not less than every 4 feet with metal or fiber mesh strapping. There are no ducts through the fireplace chase.
5. Flex ducts are secured so to not crimp or crush the duct system. The duct's R8 insulation is not effective when crushed, and crushing can affect air flow.
6. The structural integrity and dimensions of the platform and catwalk meet the HVAC codes and HVAC contractor's access needs. If not, contact the framer to revise the platform as needed.
7. The system's refrigerant and power lines to the outside compressor from the house were installed and wrapped, and the house is ready for setting the compressor at a later stage.
8. Roof vents of all kinds were installed through the roof at the back or sides of the home only, not on the front elevation of the home.
9. All non-HVAC system venting, such as kitchen hoods, bath vents, and dryer vents, was installed through the roof decking.
    - Note: Long runs on dryer vents of more than 30 feet require a booster blower fan per code; avoid if possible.
    - Bends create friction; add footage for turns:
        - Per each 90-degree turn, add 5 feet.
        - Per each 45-degree turn, add 2.5 feet.
10. Furnace flue pipe is strapped every 4 feet, and the nail heads on the strapping are sealed with roofing sealant.
11. The work area in the home and attic is clean, and all debris was placed in the designated receptacle.

STAGE 6: MEP ROUGH-IN STAGE

ELECTRICAL ROUGH-IN

Timing: The home is ready for electrical rough after the HVAC rough-in is complete.

Tasks:

- The electrical contractor supplies all material and labor required to install all wiring to all switches, plugs, light fixture locations, and appliances per the plan.
- Contractor connects the wiring to the air handler and furnace already in the attic.

Note: Electrical contractor provides all material and labor.

Confirm the following:

1. Wiring to all switch, plug, light fixture, whirlpool tub, and appliance locations conform to the electrical plan and include all buyer options.
2. No switches are located behind doors; note door swings.
3. Plug heights (usually 18 inches) and switches heights (usually 54 inches) were installed per specifications and are the same height in any one room. Aging-in-place options may change heights; confirm.
4. No wiring runs perpendicular to joists except on and along strong-backs, and those are securely strapped. No wiring is installed in fire stops or return airs or run across mechanical platforms or other walkways.
5. There are no splices, crimps, or twisted wire.
6. Electrical "homeruns" were drilled through the top plate, not around it.
7. Outside plugs are installed on separate circuit, ground-fault circuit interrupter (GFCI). There are a minimum of two GFCI plugs in the primary bath. Other GFCI locations are per plan and local code.
8. Exterior light fixtures are wired to the correct location and height per plan.
9. Any openings or damage to sheathing by the electrician are repaired with appropriate tape material.

10. All bedrooms are blocked and wired for fans per the plan.
11. Vanity area wall light heights conform to the interior elevation and specifications with the mirror plan.
12. Directional can locations for centered lights, such as for over the fireplace, are centered appropriately per the plan.
13. Nail guards are installed wherever wire is not drilled through center of the stud. More nail guards are better than too few.
14. Electrical wiring is pulled through to floor plugs and the island.
15. Gas pipes are grounded per code.
16. Power was installed to the furnace and the air handler in the attic.
17. Ceiling fixture and recessed cans were installed (or wired only) in all locations per the electrical plan. If framing members conflict with fixture locations, design a solution and confirm with buyer. This is not a problem when using LED can lights.
18. House is swept and clean, and all debris was placed in the designated receptacle.

STAGE 6: MEP ROUGH-IN STAGE

ELECTRICAL UNDERGROUND

Timing: Builder should schedule the electrical underground as soon as the electric rough-in is complete and the meter base location is determined, either before brick is installed or immediately after the brick is cleaned.

Tasks:

- Electrical contractor trenches between the transformer and the service entrance and meter location on the house.
- The service wire is placed in a PVC conduit and installed in the ditch connecting the house to the transformer. Service wire should be inspected at this time.
- After installation, the trench is backfilled/compacted to the level and compaction of the surrounding soils.
- Contractor makes the required connections at the meter base, or wraps the wire ends if the meter base is not installed.

Notes:

- The contractor provides all material and labor.
- With the electrical undergrounds complete, the builder may, depending on the service provider, be able to order the permanent electrical meter and service.

Confirm the following:

1. The trench is a minimum of at least 27 inches deep; this may vary by jurisdiction.
2. Trench is backfilled and recompacted level.
3. Tunneling under sidewalks is common; confirm there is no damage to walks.
4. The PVC riser at the electrical service location is plumb, and there is no damage to the exterior cladding.
5. The trench location is properly flagged near any future fence location.

STAGE 6: MEP ROUGH-IN STAGE

LOW-VOLTAGE WIRING ROUGH-IN

Timing: The home is ready for low-voltage wiring when the electrical rough-in is complete.

Tasks:

- Builder reviews the community standards and all buyer-selected options with the installing contractor prior to starting the work.
- Contractor installs the low-voltage wiring, including all components of the security system, audio visual wiring, and any optioned structured, data line wiring.

Note: The low-voltage contractor may be a low-voltage specialty company, or in some communities, the contractor may be the electrical contractor.

Confirm the following:

1. Contractor and builder are clear on all standard keypad(s), master control boxes, smart switches and plugs, speaker locations, and all other locations or other options selected by the buyer.
2. A 110-volt plug is located within 12 inches of the security system's master control panel location. Usually these are in the master closet or a coat closet; confirm no shelving or shoe racks obstruct the control box location.
3. Attic access openings and walkways in attic are not obstructed or crossed by any wiring.
4. Security sensors were installed on every door and window opening.
5. Security sensors are placed on the sides of the windows only. Never penetrate the window pan flashing to avoid water entering the inside of the wall.
6. Data line wiring was installed as designed:
   a. All options and locations selected are in place.
   b. Wiring is never crimped, folded, nor stapled into a 90-degree angle.

   c. Having only one drop per run, there are nc daisy chains (homeruns only).
   d. No low-voltage wires run parallel to high-voltage (110) wiring to avoid interference.
7. Cable and phone outlet locations have 110-volt service within 24 inches of each outlet.

STAGE 6: MEP ROUGH-IN STAGE

INTERIOR CLEAN: VACUUM FRAME

Timing: After all mechanical rough-ins, frame punch, the rough clean, and preferably after all inspections have passed and prior to insulation, the frame of the home must be thoroughly vacuumed.

Tasks: The clean contractor sweeps the slab completely and vacuums all frame cavity walls free of all wood shavings and all debris of any kind remaining in the home at this stage.

Note: This clean is especially important to your buyer and sets the stage for your buyer's precover walk and a home free of past construction debris.

Confirm the following:

1. Top and bottom plates and fire-blocking are clean, and there are no drilling chips left.
2. There is no micro trash or food, or other organics left in the walls or anywhere in the house.
3. Return air chases are clean.
4. Dead spaces under stairs are clean.
5. Attic catwalk was swept clean.
6. Garage was swept clean.
7. Inside and around tubs and showers are clean.
8. All trash was placed in receptacles, and the house is clean.

STAGE 6: MEP ROUGH-IN STAGE

PRECOVER INSPECTIONS

Timing: When all individual plumbing, electrical, and HVAC mechanical inspections are passed, the house is ready for a series of other precover inspections.

Tasks:

- Inspections at this stage include the city inspector, the project manager and/or the company's third-party inspector, and the buyer and possibly the buyer's third-party inspectors.
- All parties must sign off that all structural, mechanical, and electrical component was installed in accordance with all codes and that all planned features and options are complete and in their correct locations.

Notes:

- Passing the precover inspections is required for additional work to progress from this point.
- This step of critical inspections puts many eyes on the work performed thus far, allows for correction as needed, and confirms that the house is ready to cover the framed walls.
- When the home is sold at this stage, the builder schedules and conducts a buyer precover walk. This walk is optional to the buyer, but it is important in building the client's confidence going forward and in obtaining signoff now for the work installed, thus avoiding any "Sorry, it's too late now" discussions later. See the buyer precover walk section for more details.

Confirm the following:

1. Collect, organize, and have ready for any inspecting party all relevant, job-specific documents such as form survey, cable stress certificate, any third-party inspection

reports, and city-stamped sets of architectural and engineering plans and permit forms with all passed MEP inspection documents and more.

2. Property address and identification are highly visible, and any required permit or safety postings are in place.
3. Any of the buyer's specially requested repairs or concerns previously discussed were addressed and can be demonstrated to the buyer during the buyer pre-cover walk. Have all correspondence on these issues on hand for walks with any inspecting party, but certainly for the walk with the buyer.
4. All incidental damage (e.g., holes in sheathing, broken windows, shingle damage) to the home is repaired prior to the precover inspections.
5. Framing members affected by drilling or notching during the rough-ins are in good and acceptable condition, and all bowing studs were repaired or replaced.
6. The front yard and backyard, the garage, and everything about the home was swept and is clean with no debris anywhere.
7. All cable grouting is still intact, and all concrete honeycombing and damage was repaired.

MOST FAILED PRECOVER INSPECTIONS ITEMS:

- No or low gas test
- Tubs/showers not water tested
- Smoke detectors missing or too close to registers/return air
- Return air plenum not completely sealed
- Missing plumbing or electrical nail guards
- Third-party documents, truss layouts, and other documents missing
- Framing contractor–related:
  - No repair letter on field-cut trusses
  - Bottom plate improperly anchored
  - Windstorm strapping, tie-downs, and bracing incomplete
  - Sheathing not nailed vertically and at every stud
  - Ceiling, floor joists, or rafters not per engineered drawings
  - Strongbacks improperly installed or missing

- » Corner and wall bracing incomplete
- » Missing or cavities/holes in firestopping
- » Beams not installed per plan or not fully bolted
- » Improperly sized, improperly nailed, or missing joist hangers
- » Oversized holes in studs, joists, or other framing members
- » Over notched bowing stud repairs
- » Improper splicing and bracing of hips, valleys, and ridges

8. Safety railings at the staircase and at the back and front porches are in place and sturdy.
9. Fireplace installation is sturdy and strapped, a l required f restop is in place, and the chimney cap is installed and level.
10. The home is swept and clean, all erosion control is in good condition, and all signage is in place.

STAGE 6: MEP ROUGH-IN STAGE

BUYER PRECOVER WALK

Optional Precover Walk

Timing: Builder schedules the buyer precover walk to occur during the precover inspection phase. It is preferable for the walk to occur after passing the city precover inspection because at that point the builder can give the buyer a more accurate forward-looking construction schedule and that fact will build the buyer's confidence in the home.

Task: This is a home demonstration and is not a frame/mechanical punchout. It is an opportunity for the builder to confirm the selected options are installed as well as to enthusiastically show off the home to the buyer.

Notes:

- Builder should take the lead at the onset of the meeting and present the home's engineered skeletal construction and its built-in MEP features and control box locations.
- Be prepared to review the going-forward schedule of the home as of the walk date. This is the time to reset any scheduling expectations that are possibly held by the buyer.
- Builder reviews and confirms any known outstanding balances due on selections made and advises the buyer to promptly connect with their salesperson to make necessary payments needed to avoid delays in ordering and receiving the selected materials.

Confirm the following:

1. Buyer approves and acknowledges that all selected options were installed in the walls per the job's option summary report.

DRYWALL STAGE

STAGE 7

- Poly seal I
- Wall insulation
- Drywall installation and tape, float, and skim and sand walls
- Exterior cladding installation:
    - Stucco system installation
    - Brick installation
    - Stone installation
    - Independent code inspector's masonry inspection
- Garage door installation
- Clean and rough grade the lot

The drywall stage starts with the sealing of any possible points of air infiltration and installing wall insulation. While the work inside is ongoing, the installation of the exterior cladding (stone-brick-stucco) is underway. After insulation but not before stucco, the drywall may be installed. With the drywall up, the garage door may be installed. And as usual, the stage finishes with a lot clean and a rough grade to reestablish positive drainage once again around the house and to the street according to the plot plan.

STAGE 7: DRYWALL STAGE

POLY-SEAL I

Timing: When all pre-drywall inspections are passed, all punchout is complete, and immediately before insulation and drywall installation

Tasks:

- Poly-seal I includes installing baffling for blown insulation, caulking, and foam sealing around doors, windows, and any exterior wall penetration as needed to reseal the entire building envelope.
- The building envelope must be completely sealed at all the exterior seams, edges, and framing connections.

Notes:

- Poly-seal may or may not be a two-stage process. Either process works, but a poly-seal II trip by this contractor is likely.
- Only low-expansion foam is placed around windows.
- Fire-rated foam may be required in certain locations; confirm local codes.
- Foam sealing is to be done in accordance with the company's energy program requirements.

Confirm the following:

1. Baffles are installed continuously along and between every rafter, running from the soffit ventilation along the decking to reach above the level of the blown insulation, thus allowing air to reach from the soffit to the attic space (no cardboard baffles allowed).
2. Cavities around all doors are sealed. Large gaps may be filled with sheathing wedged into gaps as needed.
3. Low-expansion foam applied around windows. Close and lock all windows when injecting the foam into the cavities around the windows.

4. All exterior wall corners are sealed from floor to the attic exterior plate and around all horizontal fire blocking sealed top and bottom.
5. All openings made by wires, electrical boxes/switches, and/or plumbing pipes are sealed at top plate and any other exterior wall penetrations.
6. Overexpanded foam outside of the wall cavity is cut away and not left protruding.
7. The house is clean, and all contractor's debris is in the designated trash receptacle.

STAGE 7: DRYWALL STAGE

INSULATION: WALLS

Timing: After the poly-seal 1 is complete, the walls are clean, and all precover inspections are passed, the home is ready for wall insulation.

Task: The insulation contractor provides all materials and labor to install the approved wall insulation system in accordance with the builder's insulation standards and the contractor's proposal.

Notes: Insulation standards include:

- R-38 blown attic
- R-22 sloped ceilings/hot wall with batts
- R-15 attached garage walls with batts
- R-13 exterior walls with batts

Confirm the following:

1. All framing cavities, including boxed-out windows, arch top windows, bay areas, cold walls under stairs, garage walls, and other irregular framing features of the home, are completely insulated.
2. Insulation was installed around recessed can fixtures and around the attic access stairs.
3. Batt insulation was installed (stuffed) at all exterior corners or other framed areas where framing is too tight for blown insulation to reach.
4. No batts are overstuffed or crushed. Overstuffed or crushed insulation does not function to specification.
5. Insulation was installed on the outside of each water line on an outside wall.
6. Insulation was installed below the heating, ventilation, and air conditioning (HVAC) platform in the attic.

7. Confirm batts are thoroughly stapled, at least 12 inches on center at every framing member.
8. Confirm 100 percent installation of batts in vaulted, gambrel ceilings cathedral, and other rooms partially extending from the main house where access prevents blowing the ceiling.

STAGE 7: DRYWALL STAGE

DRYWALL

Timing: The home is ready for drywall installation when all precover inspections are passed and wall insulation and poly-seal I are complete. Drywall is always installed prior to exterior stucco.

Tasks: The drywall contractor hangs all drywall and installs metal corners, tapes, and floats and skims all corners, joints, and nail heads. The contractor textures the ceiling and spray-textures all walls per the specifications. Cement-backer board is installed in all bath wet areas (i.e., tubs and showers).

Notes:

- **Drywall specifications:**
  - » 4 feet by 12 feet by 0.5 inches first floor
  - » 4 feet by 12 feet by 5/8 inches fire code–designated areas (i.e., garage)
  - » 4 feet by 10 feet by 0.5 inches second floor
  - » 3 feet by 5 feet by 0.5 inches cement-backer board for wet areas
- **Nailing specifications:**
  - » Wall application: one nail on both outside edges and three or more nails in the field
  - » Ceiling application: one nail on both outside edges and four or more nails in the field

Confirm the following:

1. Drywall delivered to second floor is not stacked taller than 20 sheets.
2. There is no viably bowed drywall on walls or ceiling. Raised or decorative ceiling lines need to be visibly straight and plumb to the eye.

3. All vertical drywall joints are staggered.
4. Outside cornerbead was installed using the specified style bead (square, small rounded, and wide rounded) and secured with drywall nails at 8 inches on center (OC).
5. Cement backer-board (3 feet by 5 feet by 0.5 inches) is installed with corrosion-resistant fasteners in all wet areas.
6. All plugs, switches, recessed cans and light fixtures, medicine cabinets, and attic ladder access openings are cut out and visible.
7. There are no nail pops, no ripples in tape, and no ridges visible after skimming. Walls are smooth to the eye when one is standing in the center of the room.
8. Finished outside edges at cased openings, window returns, and the area around plugs and switches are cosmetically excellent, and window returns are rocked fully to the bottom of window.
9. There are no chips or other damage to tubs or shower pans, and no nails remain in the tub or shower pan.
10. All HVAC supply, return, and vents and washing machine, gas valves, fireplace, windows, and ice maker boxes and recessed cans are properly covered and protected prior to application of the texture.
11. Texture is uniform and even to the eye while standing in the center of the room, and inside corners are not visibly double-textured.
12. All closets are completely textured.
13. Burn marks from over sanding are reskimmed and sanded.
14. The garage is taped, floated, and textured the same as the house.
15. The floor is scraped clean of all joint and compound materials, and the house is swept clean, with all debris placed in the designated location.

STAGE 7: DRYWALL STAGE

EXTERIOR CLADDING THE HOUSE: STUCCO SYSTEM

Timing: Stucco homes must wait until after the drywall is installed to install the stucco cladding because installing drywall can damage stucco.

Tasks:

- Builder and the stucco contractor inspect the cornice for any incomplete or damaged areas, paying special attention to the flashing installation locations and condition. Any flashing concern must be corrected prior to applying the stucco.
- The stucco application involves two phases: the oriented strand board (OSB) sheathing housewrap phase and the stucco exterior system phase. Contractor applies the OSB sheathing and housewrap and exterior stucco system per the home's exterior elevation, following the plan details exactly.

Notes:

- Builder provides the OSB sheathing and housewrap. The stucco contractor provides all materials, equipment, and labor to apply the stucco system.
- Scaffolding must meet OSHA (Occupational Safety and Health Administration) standards, standing on solid ground or solid platforms of wood and be in good condition and 100 percent complete, including toe kicks.
- The stucco application can take up to 14 days to properly cure and dry and thus avoid cracking. Contractor must consult with builder before adding any curing additives.

Stucco process in summary:

1. OSB sheathing is installed at all stucco areas.
2. The house is wrapped using the approved housewrap product.
3. Lath is fastened to cover the walls over the OSB and housewrap.

4. Scratch coat (scratch and brown cement mixture) is applied to the lath.
5. Sand finish coat (sand/lime Portland and masonry cement mix) is applied.
6. Sanded primer coat is applied to the entire wall area.
7. Elastomeric finish coat is troweled on the walls completely.

STAGE 7: DRYWALL STAGE

SHEATHING, HOUSEWRAP, WINDOW INSTALLATION, AND STUCCO

Confirm the following:

1. OSB 4 feet by-8 feet sheathing is installed horizontally and nailed 6 inches OC and using H clips along all horizontal and vertical joints.
2. OSB sheathing is installed vertically 4 feet out from all inside and outside corners.
3. Holes cut for plumbing are not greater than 0.25 inches oversized over the pipe.
4. Housewrap is installed horizontally over all the OSB sheathing, overlapping more than 6 inches at each horizontal seam; seams are taped closed 100 percent with approved sealing tape.
5. Windows are installed in stucco walls by cutting a Y in the housewrap at the window's rough openings and securing the wrap with cap nails. The installed windows are taped on all four sides, extending 6 inches from the window.
6. Confirm that windows are installed perfectly square in the locked position with 3/8-inch panhead screws and completely caulked edge to edge behind top window flange.
7. Z bar flashing is installed using window tape above between the window and the OSB and under the housewrap and extending at least 0.5 inch past the sides of the window.
8. Contractor has all windows protected with 4 millimeters poly.
9. Next, lath is installed and nailed thoroughly into the OSB over the entire area to receive stucco, and all window and decorative banding is built using white foam covered with lath.
10. Confirm control joints are installed for every 100 square feet at 10-foot intervals. All windows and door have vertical control joints. Minimize control joints on the front of the home.

11. Scratch coat is applied using scratch (premix fiberglass mixture only) and brown cement mixed with torpedo sand.
12. One coat of sand finish applied and floated evenly using mix #4 sand, lime Portland, and masonry cement. (Sand must meet ASTM C144 standards.)
13. One coat of sanded primer is applied to the entire stucco wall surface and completed by applying an elastomeric finish over the entire surface.
14. Weep screen is installed and clean of mortar along the bottom of all horizontal edges.
15. Depending on the city and other considerations, the builder may choose to install a fiber cement strip 4 inches tall (7/16 inches fiber cement board), wrapping the base of the house below the OSB sheathing and attaching to the bottom plate to prevent wicking ground moisture from irrigation or blown rain.

STAGE 7: DRYWALL STAGE

EXTERIOR CLADDING THE HOUSE: MASONRY—BRICK OR STONE

Timing: After passing all precover inspections, the house is ready for brick or for stone. If the elevation shows stucco and masonry, the stucco is applied first and creates the edge to connect to the masonry; then both will wait until after drywall.

Tasks:

- The masonry contractor installs brick and/or stone per the plan's exterior elevation, following all plan details exactly.
- Prior to commencement of work, the builder and contractor review the elevational detail drawings and walk all sides of the house to confirm the designs and areas to receive brick or stone and the type of joint rake and mortar color to be used.

Notes:

- The masonry contractor supplies and installs all wall ties. Many municipalities require a separate brick wall-tie inspection; confirm and schedule accordingly.
- Brick or stone may be laid only when the temperature is 40 degrees and rising—no exceptions!
- *Important:* Schedule gas underground as soon as masonry is complete.

Mortar mix standards:

- 2 parts Portland cement
- 1 part lime
- 8 parts masonry sand

Confirm the following:

1. Contractor stages the brick or stone around the home, starts at the front of the home, and works around the sides, finishing in the back of the home.

2. The brick or stone is properly mixed to ensure the brick or stone's colors are evenly mixed throughout the wall. There is no concentration of any single brick or stone color in one area.
3. Only saw-cut bricks are used at the front of the home or in visible areas; no jagged broken ends are visible.
4. Brick ties are 22-gage and at 16 inches or 24 inches OC or per framing design.
5. Iron lintels were installed over all openings to support brick or stone.
6. All windows are protected and covered with poly.
7. Weep holes are installed in masonry walls every 33 inches; they must be completely clean of any mortar.
8. Expansion joints are installed in masonry walls every 25 feet.
9. Masonry corners are plumb and straight. Hang a plumb line to confirm.
10. There is no excessive mortar accumulation in the wall cavity.
11. All brick and stone and elevation details are followed exactly. Brick and stone installation and locations must follow the detail exactly.
12. Masonry window rowlocks were installed on a greater than 5 percent tilt, and the brick or stone extends below and under the frame of the window.
13. The top-course brick or stone is flush and level with brick frieze.
14. All mortar joints are tooled and clean. Vertical joints align every other course to form a visually obvious pattern.
15. All window tracks and window frames are clean of mortar, and there are no scratches or other cosmetic damage or broken windows.
16. Hose bibs are secured and at a 45-degree angle from the wall, and the valve is accessible to grip with the hand and flush with the masonry wall.
17. Plumbing cleanouts, gas lines, electrical plugs, insect protection boxes, or other protrusions to the wall are secure, plumb, square, fitting flush with masonry wall, and not recessed into the brick.
18. The precast address block was installed in the location per plan.

19. All exterior light fixtures, coach lights, flood lights, entry lights and exterior plugs, and the doorbell are properly positioned at the designated height per plan, and the boxes are flush to the brick.
20. All broken bricks and mortar were scrapped to the designated location.
21. All unbroken brick or remaining stone after completion was stacked for the homeowner in the garage. Builder should collect a minimum of 20 loose bricks or stones to save for the homeowner's future repair needs.
22. Upon completion or according to the timing required by the municipality, the masonry contractor arranges for the masonry wall-tie inspection, if required.

STAGE 7: DRYWALL STAGE

INSTALL GARAGE DOORS

Timing: Garage doors may be installed when the drywall installation is complete and before the interior trim stage.

Tasks: The contractor installs the garage door tracks, places the horizontal garage door panels into the track, and bolts the panels together to form the completed door. Next, the springs are installed, and the tension adjusted to allow the door to open, close, and lock easily and smoothly.

Note: The garage door contractor provides all material and labor.

Confirm the following:

1. The door pattern matches the pattern shown on the elevation or as specified by the buyer selections summary.
2. The garage door(s) opens, closes, and locks easily.
3. The tracks are firmly secured to the solid framing members.
4. In cases when the garage is dropped, the contractor installs the required extended mounting components to fit as built the ceiling height.
5. All railing and trim components at sides and top were installed secure to the framing members.
6. The top of the garage door is level with the opening's header.
7. When closed, the entire door rests flush to the wall on all three sides, with no light from the top or either side.
8. There is no damage to drywall.
9. The garage door keys work properly and are secured in the job file.

STAGE 7: DRYWALL STAGE

TRACTOR: GRADE AND CLEAN AFTER BRICK AND DRYWALL

Timing: When the brick/stone application and the drywall is complete, the house is swept clean, and all scrap material is placed in the designated location.

Tasks:

- The grade-and-clean stage is performed to clean drywall and masonry debris. Clean contractor hauls off all trash on the lot, including broken brick, mortar, concrete overspill, stone, drywall, and other debris associated with stage 7 activities.
- Clean contractor places drywall scrap in the designated location in a single pile, typically in a single pile staged less than 10 feet from the curb.

Notes:

- With the brick complete, the builder may now hang shutters or other exterior ornamentation, if any.
- Reestablish erosion control as needed.

Confirm the following:

1. The lot is completely clean of all building debris, including micro-trash.
2. The silt fence and all erosion control and signage are in place.
3. Port-O-Let, wire basket, and other trash receptacles are removed from the yard at this time in anticipation of the final grade.
4. Remaining are bricks left stacked in the garage for the homeowner; leave a minimum of 20 stones or bricks.

INTERIOR TRIM AND CABINET STAGE

- Masonry counterflashing
- Permanent electric and gas meters
- Interior trim I
- Poly-seal II
- Set cabinets (measure for tops)
- Paint wall-out and enamel
- Bath tile

The interior trim and cabinet stage is focused on finishing out the interior or the home, including installing interior trim and doors, cabinets, interior paint, and bath tile. On the exterior, the stone/brick masonry may be cleaned and completed leading up to the installation of the counterflashing and utility meters on the exterior of the home.

STAGE 8: INTERIOR TRIM AND CABINET STAGE

MASONRY COUNTERFLASHING

Timing: When the brick/stone installation and cleaning is complete, the roofing contractor returns and cuts the counterflashing into the brick in all locations where the shingles are in contact with the brick or stone veneer.

Tasks:

- Builder and contractor review the plans and house to agree on installation locations, generally along the rakes of the bricked gable walls and against the brick chimney.
- The contractor saw-cuts the mortar joint, inserts the metal flashing edge into the cut mortar joint cut, and seals the connection with butyl caulking to secure the flashing down the face of the brick/stone and lying flush on the roof deck.

Notes: Contractor must notify the builder of any flashing concerns where other measures, such as taller flashing, cricket repair, or valley pan, may be advised. Flashing is the source of almost all roof leaks, and too much flashing now is better than adding flashing after a roof leak.

Confirm the following:

1. The flashing is smooth and straight, and the condition is good.
2. Flashing is properly and completely sealed into the mortar cut.
3. Finished product looks professionally installed.
4. There is no shingle damage.
5. The L and J flashings are properly laced to keep the water on top of the shingles and roof.
6. Roof is clean; there is no mud or footprints visible from the ground.

STAGE 8: INTERIOR TRIM AND CABINET STAGE

PERMANENT ELECTRIC AND GAS METERS

Timing: With exterior cladding, masonry, and/or siding complete, the builder may submit the requests for both the permanent electric meter and the gas meter.

Tasks:

- Remove the T-pole and set the permanent residential electric meter at the house.
- Contact the gas company and request that the home's gas meter be installed by that provider.
- Complete and pass all required inspections to secure green tags.

Note: Confirm the specific process for the jurisdiction and municipality for the job site.

STAGE 8: INTERIOR TRIM AND CABINET STAGE

INTERIOR TRIM I

Timing: After the walls and ceilings are textured, the home is ready to install the interior trim I.

Tasks:

- Interior trim I is the installation of all interior doors, base boards, stairs and stair parts, windowsill and apron, and crown moldings and the building of fireplace mantel and other decorative wooden wall moldings.
- Builder and contractor walk the home to discuss trim termination locations and door swings relative to light switches and other trim I locational questions.

Notes: The trim contractor supplies the nails and labor, and the builder supplies all the trim materials. See the purchase orders for component quantities and details. Some communities may use turnkey trim contractors; confirm for your community.

Confirm the following:

1. All doors are installed plumb and level and blocked with wood shims, all doors close flush against the jamb on all sides to top, and both sides have a reveal of 1/8 inch and a bottom reveal of approximately 1 inch, depending on flooring thickness.
2. All door casings and trim connections are cut neatly and tight at 45 degrees, and all hinge pins are set.
3. Baseboard was installed level and is raised 3/8 inch above the subfloor.
4. Window stool is shimmed level, routed on three sides, and projecting 1.25 inches left and right and installed on all windows that are visible from any floor.
5. Wooden closets rods and shelving were installed to match the plan layout. Shelves were installed level and cleated continuously below all shelves (including returns) and vertically along every shelf every 36 inches.

6. The stair system was built per plan and code:

   a. Contractor removes the temporary stair railing as needed to build the permanent stair railing and secures all the reusable safety brackets for the builder.

   b. Handrails are continuous to and from the landing at the top to the first step at the bottom of the stairs.

   c. The wall-rail gap is never narrower than 1.25 inches between wall and rail.

   d. Stair treads have a tread depth of not less than 10 inches. Risers are not greater than 7.75 inches tall, and there is no variation in risers more than 3/8 inches.

   e. Newel posts were installed every 8 feet of horizontal handrail and anchored into structural faming.

   f. Stair spindles/balusters are bored into the handrail and into the base cap and then glued and nailed.

   g. Stair spindles/balusters spacing meets International Residential Code (IRC): maximum spacing of 4-inch on center (OC).

   h. Rosettes were used at all railing terminations into a wall.

   i. Wall skirt of 1-inch-by-12-inch board was installed on both sides of the stairs on two-sided stairs or on a stepdown profile stair design.

   j. Rail height ranges between 36 and 44 inches at stair steps and horizontal runs on balconies, game rooms, and landings (IRC).

STAGE 8: INTERIOR TRIM AND CABINET STAGE

POLY-SEAL II

Timing: The house is ready for the poly-seal II once the interior trim I is complete.

Tasks: Contractor seals with caulk or other approved material along all baseboards, electrical switches and outlets, air conditioning (AC) plenum boxes, light fixtures, security control boxes, and any other penetration in the walls or ceiling.

Notes: Contractor supplies all material and labor for this task. Contractor must use an approved poly-seal material or a 25-year paintable caulk depending on the location.

Confirm the following:

1. Poly-seal was installed in all designated locations, including but not limited to every electrical fixture and switch box and low-voltage boxes, and continuously along the baseboard and any other home component that is part of forming the building envelope.
2. Home was swept, and all debris is in the designated receptacle.

STAGE 8: INTERIOR TRIM AND CABINET STAGE

SET CABINETS

Timing: The home is ready for cabinets when the drywall and texture is complete.

Tasks: Cabinet contractor delivers and installs the previously measured cabinets complete with all parts/trim for the kitchen, bath varities, and other locations per the contract, approved shop drawings, or interior elevation plan.

Notes:

- All materials and labor for the cabinets are provided by the cabinet contractor.
- As soon as the cabinets are set, schedule the tops contractor to measure all countertops.

Confirm the following:

1. Cabinets are installed with 3-inch screws to the walls 12 inches OC top and bottom of the wall cabinets and 12 inches OC at the top of base cabinets. Cabinets side by side are fastened together at the styles 12 inches OC.
2. Two 1-inch holes are cut in sink base cabinet on the dishwasher side—one at the top corner for dishwasher drain and one at the bottom corner for water line.
3. Plumbing holes cut in the back of sink base cabinets have a maximum gap around the plumbing pipe of a 0.5-inch gap from pipe to cabinet.
4. Tops of all base cabinets are level and shimmed at the back of the cabinet as needed until the cabinet tops are level. Confirm using a level.
5. Kitchen sink base cabinet is properly blocked around the sink cutout to support the still-to-be-installed sink.
6. Wall and base cabinets align properly from above and below or are centered appropriately where cabinets are centered on windows or other pass-through openings.

7. All wall cabinet moldings (crown or lip molding) and finished cabinets ends were installed per the community standards, and the contract. Trim was installed as needed to cover all base and wall cabinet-to-wall gaps.
8. All protruding screws, nails, and staples inside cabinets are cut clean.
9. All cabinet doors are level, align with one another, and close flush to the door jamb all the way around the cabinet door. No binding or warped doors are allowed.
10. Drawers open and close properly, smoothly, and completely.
11. If soft-close cabinet door hardware was chosen, confirm that it is installed.
12. All adjustable shelves are logical and uniformly hung inside all cabinets. Homeowner should not have to readjust shelves to move in.

STAGE 8: INTERIOR TRIM AND CABINET STAGE

WALL-OUT AND ENAMEL

Timing: The house is ready for wall-out and enamel when trim I is complete, and all walls are repaired as needed and the texture is completely dry.

Tasks: Depending on house size, wall-out and enamel is at least a three-day job performed in the following sequence:

Day one:

- The painter first thoroughly sweeps the house and cleans all paint surfaces as needed, then preps by caulking all trim to walls and all wood joints and connections.
- Painter covers and protects all not-to-be painted surfaces (cabinets and other prefinished surfaces) using plastic or other protective measures taken as agreed upon by the builder.
- Painter sprays a lacquer undercoat on all doors, door trim, window stools/tops, and bottoms of doors.

Day two:

- The painter sprays the walls and ceiling of the house and the garage with the specified interior wall paint. One coat is to be applied to all unprimed trim, base, casing, and shelving wood.
- Painter sands all doors and wooden rough edges, kerfs, etc., as needed and then sprays enamel trim paint on all doors, trim, cleat stock, shelving, pole sockets, and any decorative moldings.

Day three:

- Painter cuts-in all enamel overspray around doors and windows using wall paint to create a professional interior paint finish and then rehangs all doors as needed.

Notes:

- Care must be taken in choosing the location for spraying doors; avoid all non-carpet areas.
- Painter must take special care in cleaning or disposing of paint and paint-related chemicals.

Confirm the following:

1. Paint contractor is using only approved paint and caulk products.
2. All wood joints or connections are caulked completely, creating a smooth transition at the connection. Preparation is the key to a good paint job.
3. All doors are rehung, and all hinges and door parts are replaced such that the door works perfectly.
4. Stair parts and other adjacent components are properly protected, and no overspray remains after completion.
5. In noncarpeted areas (i.e., wood or tile areas), the floors are protected to avoid overspray on the slab. Glued-down wood floors and tile adhesive can be affected by overspray on the slab.
6. Overspray protection used at door hinges and at plumbing fixtures such as bathtubs and shower pans are protected, and no overspray remains.
7. There is black paint inside all jump ducks and air supply boxes.
8. There are no runs or slick/shiny spots from enamel overspray.
9. The pantry, coat closet, and bedroom closets were painted completely, and all closet shelves were caulked to the wall and painted on all sides.
10. All tops and bottoms of doors and door trim are painted completely.
11. All empty caulk tubes, paint cans, and other debris was removed from the home and placed in the designated receptacle.

STAGE 8: INTERIOR TRIM AND CABINET STAGE

TILE: BATH TILE

Timing: The home is ready for wall and floor bath tile when the wall-out and enamel and possibly the cabinets and countertops are complete. The tile contractor supplies all material and labor for this work.

Tasks:

- Contractor sets tile over the cement backer board in tub and shower wall surrounds and installs all wall and bath floor tile per the plan, the buyer selection summary, and any provided shop drawings.
- After tile is set up, joints are grouted using the selected color, and the tile wall is cleaned.

Note: An antifracture membrane may be installed below the floor tile as a standard or as an upgrade, depending on the community specification, which is a function of the plasticity index of soils in the community.

Confirm the following:

1. Tile/grout color and pattern delivered match the buyer selections summary. Contact sales or the design center if items do not match.
2. Floors are properly prepped and floated to be level using an approved floor leveling system.
3. Tile joints and surfaces are level one tile to another, forming walls that are smooth with no wavy or bowed areas visible.
4. Grout was installed with no joints exceeding 1/8 inch, with no cracking.
5. Soap dish and/or towel bar per plan and inside corners were caulked with clear silicone.

6. If there are steps or a window in the tub/shower area, confirm all horizontal surfaces have proper slope (minimum ½ inch) away from the window or wall and toward the tub/shower area.
7. All woodwork and walls are clean of grout, and the work area was swept, with all debris placed in the designated receptacle.
8. There is no tub or shower pan damage.
9. Tile contractor must take special precautions with cleanup to avoid improper disposal of grout dyes and other adhesives used in their tasks.

FINAL GRADE STAGE

STAGE
9

- Blown attic insulation
- Exterior topcoat paint
- Set countertops
- Tile: backsplash, floors, and fireplace
- Interior trim II
- Reglaze windows
- Install gutters
- Tractor cut driveway and walks
- Flatwork: form and pour
- Lot clean and final grade
- Install front yard trees

The final grade stage is a mixture of tasks occurring both inside and out. With the cabinets installed, the natural progression is to the countertops and backsplash installation and any tile floor areas. The ceiling insulation can now be blown, and it's time to reglaze any broken windows. Outside, the work to preserve the established grading is underway, with the installation of the flatwork and gutters and the final grading of the lot. With the final grade in place, it is time to plant any community-required yard trees.

STAGE 9: FINAL GRADE STAGE

BLOWN ATTIC INSULATION

Timing: The house is ready for attic insulation after the house is drywalled and the exterior cladding is complete and after electric hot punch (in case of repairs).

Task: Contractor blows the specified insulation a minimum of 18 inches deep throughout all designated flat ceiling areas of the attic.

Notes:

- All material and labor for this work is provided by the insulation contractor.
- The layer of blown insulation is designed to achieve an R-38 blanket over the home's living area.
- Blown fiberglass and cellulose R Value varies from R2.2 to R3.7, based on specification.

Confirm the following:

1. The depth of blown insulation is consistent throughout the house attic. See if insulation reaches to the top of the 18-inch insulation dams installed by the framer or see the height on the installed cardboard measuring sticks.
2. All baffles remain in their correct positions between every rafter to perform as designed. Repair or replace as needed prior to blowing the insulation.
3. Recessed cans must have 6-inch recess from blown insulation. If LED cans are used, then no offset is required.
4. Inspect hanging duct work and surrounding walls for any obvious damage.
5. The signed and dated insulation certificate is stapled to the rafter above the attic opening; secure for the permanent house file.
6. Empty insulation bags are placed in the designated receptacle.

STAGE 9: FINAL GRADE STAGE

EXTERIOR TOPCOAT PAINT

Timing: The house is ready for the exterior topcoat paint once the garage doors are installed, all exterior cladding is complete, and all other exterior architectural elements are installed.

Tasks:

- Painting contractor thoroughly prepares the house prior to the topcoat paint. The contractor thoroughly putties all nail holes/recesses and caulks all material joints.
- Contractor paints the final exterior topcoat of the house using the specified paint and following the prescribed exterior paint schedule.
- The paint contractor spray paints the exterior trim first and then cuts in around the trim with the base wall topcoat.
- If the final front door is already installed, the door is stained or painted at this stage.

Notes:

- The paint subcontractor supplies all materials and labor for this task. Builder inspects all prep work prior to contractor applying any paint. The preparation phase determines the quality and performance of the topcoat paint.
- On very windy days, this task could be delayed, or special care must be taken to avoid damage to neighboring properties or vehicles. Review any cars parked in the street near the painting and move vehicles as required.

Exterior paint schedule:

- Base: siding and garage doors
- Trim: facia, garage trim, door and window trim, and louver vents
- Accent: shutters and other
- Other: roof jacks match roof, lintels are black, and flashing matches brick or roof

Confirm the following:

1. At onset of prep work, all surfaces are clean and ready to receive paint.
2. Caulk conforms to the specified elastomeric caulk type or clear silicone caulking where appropriate.
    - » Caulking schedule:
        - › Brick frieze to soffit
        - › All splices in facia
        - › All brick to trim/frieze connections
        - › All 45-degree connections
        - › Completely around all windows
        - › All brick/stone expansion joints fully caulked with masonry caulk and all protrusions through brick (hose bibs) are sealed with masonry caulk as well
        - › Exterior light fixtures and doorbell buttons caulked on top and both sides with clear silicone
3. All rough edges and saw kerfs at wood joints are sanded down and caulked and smooth.
4. All joints, dissimilar material connections, and windows are caulked completely on all sides, top and bottom.
5. All paints match the selection sheet. All paint is the approved 100 percent acrylic paint.
6. Topcoat paint was installed a minimum of 3 millimeters thick when dry.
7. There are no runs or overspray on shingles, brick, or stone.
8. Any protruding framing detail, such as a box window, is painted completely on its underside.
9. All flashing and brick lintels are painted with black oil-based enamel.
10. Painter cleans painting equipment at the designated hose bib only.
11. The job site is clean, and debris was placed in its designated receptacle. Special attention must be given to disposal of paint and related chemicals.

STAGE 9: FINAL GRADE STAGE

COUNTERTOPS

Timing: Countertops may be installed as soon at the cabinets are set and punched out.

Tasks: Countertop contractor delivers and installs the shop-fabricated kitchen and bathroom tops.

Notes:

- Contractor supplies and installs the kitchen and vanity sinks. Builder should confirm all color selections and cooktop cutout dimensions with the contractor.
- If the plumber provides the sinks, confirm sink cutout sizes with the countertop contractor at the time of measurement.

Confirm the following:

1. Depending on the size of the top area and thickness of the countertop, an oriented strand board under-decking may be used. In these cases, the decking seams must never align directly with the countertop seam.
2. Tops are leveled by shims and never back-pitched. If tops require more than a 0.25-inch shim, then the solution is to relevel the cabinets. Builder must confirm levelness of cabinets and tops early in the process.
3. Bar tops extend up to a maximum of 12 inches beyond the cabinet before needing additional brackets or bar top supports.
4. Where the tops side to a wall, the connection is caulked to tops with siliconized caulk.
5. If walls are out of square, there is no notching of walls allowed.
6. All top seams are level and smooth.
7. Vanity tops were secured to cabinets with approved adhesive.
8. All tops were wrapped or otherwise protected upon completion; there are no chips or damage at the time of installation.
9. The work areas were swept, with all debris in the designated receptacle.

STAGE 9: FINAL GRADE STAGE

TILE—BACKSPLASH

Timing: The home is ready for backsplash immediately after the countertops are installed.

Task: The tile contractor installs the kitchen, bar, or other designated countertop backsplash following the job-specific selections and design.

Notes:

- The contractor provides all material and labor for this work.
- The key to success with the backsplash is attention to detail. The selections detail provides:
    - » Tile identification numbers
    - » Grout color number
    - » Shop drawing with locations of each tile by number
- The tile pattern is identified by name, "Basket Weave" or "Subway," and the layout/pattern is detailed by a drawing.
- If there is any question or issue, such as electrical plug locations, do not guess; contact the design center. Plan how the outlets will affect the pattern and draw the pattern on the wall as needed.

Confirm the following:

1. All outlets in backsplash area are a uniform height and conform to your splash plan.
2. Walls are flat and smooth to receive tile; no bowing.
3. Countertops are clean and still protected and undamaged.
4. Tile backsplash joints should not exceed 1/8 inch for 4-inch tile and 3/8 inch for 6-inch and larger tiles. Specialty, 1-inch, and stone tiles have joint sizes specified in the drawn splash detail.

5. Start with whole tiles at the countertop and work up with partial tiles at the top row only.
6. Tile must be cut to within 0.25 inch of any electrical box, and no screw holes are covered or plugged.
7. All tile joints are horizontally level and vertically plumb.
8. Tiles are flush to each other, forming a smooth and even surface.
9. Grout was cleaned completely from the face of all tiles and from any woodwork, countertops, or other surfaces before it dries.
10. The home was swept, and all debris was removed and placed in designated receptacle.

STAGE 9: FINAL GRADE STAGE

TILE: FLOOR AND FIREPLACE TILE

Timing: Once the cabinets, tops, and backsplash are complete, the floor and fireplace tile may be installed.

Tasks:

Tile contractor lays out tile locations and pattern/direction prior to starting and confirms with the builder that the direction and pattern is correct.

Tile contractor preps and floats the floor level and installs the tile floor according to the job's selections and plan's locations.

Notes:

- Builder and contractor must confirm all tile numbers are correct per the job's buyer selections sheet and that the tile direction is running per any shop drawings or company standard.
- Contractor supplies all materials and labor for this task.
- The layout should minimize tile cuts of less than a half-tile. All cut tiles should be on back walls or less conspicuous locations.
- Fireplace surround tile is run so that any patterns or lines are run horizontally. Same tile lines are maintained across the entire mantle area.

Confirm the following:

1. Contractor has sufficiently prepared and floated the slab as needed for a level installation.
2. Tile is laid over the approved felt/fiberglass underlayment to protect against cracking under normal slab movement.
3. All joint lines are exactly parallel or perpendicular with walls and run evenly, and the joints are aligned and parallel across the entire floor.

4. Only whole tiles are used in the most visible locations; cut tiles land on back walls or furniture walls, never in the entry or along open public halls.
5. Interior doors removed for tile installation are replaced by tile contractor when tile is dry and complete.
6. Moldings removed in the process are temporarily attached to the wall near the location where they were removed, for replacement by others later.
7. All work areas were swept, and all debris was placed in the designated receptacle.
8. Contractor has erected barricades as needed to prevent people walking on the tile before it is dry.
9. Tiles remaining at the end of the job are placed in the attic for future use by the homeowner.

STAGE 9: FINAL GRADE STAGE

INTERIOR TRIM II

Timing: The house is ready for trim II after the wall-out and enamel paint is complete.

Tasks:

- Trim II includes weatherstripping exterior doors, installing shoe mold, wooden closet rod, door and cabinet knobs, door stops, and all bath hardware.
- If decorative hinges are used, they are replaced at this time.

Notes:

- This work may be done by a piecemeal or turnkey contractor depending on the contract in the community.
- Shelving installation specifications:
  - » Using metal rod/shelf system the supports are installed at a maximum of 48 inches on center (OC) using 3-inch screws in blocking or studs.
  - » Using wooden rod/shelf system, rods are cut to length on site and installed to the painted brackets at both ends of the rod.

Confirm the following:

1. All bath hardware was installed in locations per plan. Place all towel bars at 52 inches and all paper holders at 26 inches high and 8 inches in front of commode.
2. Doorstops were installed in walls behind each door or on hinges for doors not backing to a wall.
3. Shoe mold was installed along the baseboards to overhang tile locations.
4. Cabinet door hardware was installed on doors/drawers at agreed-upon locations.
5. All doorknobs are screwed in completely and operating properly. Lever-styled knobs are oriented up or down but waving in a consistent direction.
6. All door keepers are properly adjusted, and all doors easily snap closed.

STAGE 9: FINAL GRADE STAGE

REGLAZE WINDOWS

Timing: After all significant construction is completed and as the job glides into its final stages, the builder may schedule the glass and mirror contractor to remove and replace any broken window glass.

Tasks: Glass contractor inspects all windows for damaged glass, replaces any broken window glass, and cleans the window of any glazing bead or other debris from this work.

Note: If there are any back charges associated with the broken windows, the invoice for that repair may be billed to the responsible party at this time.

Confirm the following:

1. All windows are clean and open, close, and lock perfectly.
2. Inspect all windows carefully for any slightly visible glass cracks.

STAGE 9: FINAL GRADE STAGE

GUTTERS

Timing: Gutters may be installed any time after exterior topcoat paint.

Tasks: Gutter contractor supplies and installs all gutters according to the house plan or community guidelines and provides and installs splash blocks for each downspout.

Notes:

- Gutter contactor supplies all materials and labor for this work. Prior to commencing the work, confirm all locations with contractor.
- Veterans Administration loan houses require gutters on all sides, without regard to community specifications.

Gutter specifications: seamless 6inch aluminum gutters and downspouts with baked enamel finish

Confirm the following:

1. All gutters and downspouts are securely attached to the facia and/or wall.
2. Gutters slope toward the downspout. Long gutter runs may appear cosmetically odd as they drop along the facia lines but must slope down and toward the downspout.
3. All 45-degree mitered gutter seams are thoroughly caulked with approved caulk for gutters.
4. Splash guards and diverters are installed at heavy-flow locations.
5. Downspouts are plumb and extend not beyond the top of slab to allow for sod and splash block placement.
6. Downspouts discharging onto another roof area are aligned at an angle to spread the discharge across the roof surface and not concentrate the flow in on a set of shingles or a single location.

STAGE 9: FINAL GRADE STAGE

TRACTOR: CUT DRIVES AND WALKS

Timing: Once the exterior cladding, painting, and gutters are complete, the flatwork may be installed.

Task: The grading contractor cuts out where the driveway and walks are to be placed.

Notes:

The cut grade should be more than 6 inches below the street level at the edge of the driveway, and the intern should be cut wide enough to allow a 5-foot radius on the intern sweeps.

Order enough sand to ensure the job can be completed in one trip. Excess sand may be distributed across the lot.

Separately poured patio must be cut deep enough that the patio surface will be 6 inches below the slab finished floor and must slope away from the house.

The driveway approach requirements are specific to the municipality and community. Confirm and follow those specific methods; typically, the cut is 18 inches out from the back of the curb and 5 feet wider than the driveway on each side.

Driveway is cut 4 inches lower than the lip of the garage trough and 6 to 7 inches lower at the street level at the apron cut.

Confirm the following:

1. The approach is broken out at the street and curb in accordance with municipal and community guidelines.
2. There is no damage to slab, grouted cable ends, sewer cleanouts, water meter, and water lines.

3. Property pins are in place.
4. Driveway and sidewalk cut confirms to the lot unique plan designed by the builder during the starts process.
5. Street is clean and free of mud and dirt.

STAGE 9: FINAL GRADE STAGE

FORM SET AND POUR FLATWORK

Timing: Once your driveway and walks are cut to the required grade, the site is ready to set forms and then pour the flatwork.

Tasks: Concrete contractor forms, makes up, and pours the driveway, patios and approach, and sidewalks according to the plot plan.

Notes: Reconfirm all locations per the plot plan, lead walks conform to community setbacks and connections to other public walks, all easements are considered, and you have sleeved crossings for irrigation, lighting, and underground utility crossings.

Confirm the following:

1. Sand is placed in the forms to achieve finished concrete thickness of 4 inches.
2. Reinforcement 6-by-6 wire mesh and/or #3 rebar and dowls were installed per engineer's specifications, community requirements, and local codes.
3. Sleeve locations and quantities were installed for irrigation, lighting controls, etc.
4. Concrete for flatwork is 2,500 pounds per square inch (psi). No water may be added to flatwork mix.
5. Walk is a minimum of 4 feet wide and sloped with a maximum of 2.5-inch drop over the walk's 4-foot span width.
6. Walks require steps if the grade is greater than 5 percent. Add only the number of steps until the grade is less than 5 percent.

    a. Steps should be 5.5 inches tall.

    b. Steps should be broken into step groups of no more than three consecutive steps.

    c. If four or more steps are required, a 36-inch landing and rail will also be required.

    d. Score a 0.75-inch expansion joint on the inside corner of all steps.

7. Driveway forms conform to the slope plan and budgets submitted by the builder during the starts process.
8. Driveway is poured 0.75 inches lower than the apron of the garage to allow for soil movement and displacement between the garage and driveway.
9. Driveway intern flairs a minimum of 3 feet on each side at the street.
10. Standalone patios must be dropped 4 inches below the top of the foundation and sloped 0.25 inches per foot away from the house.
11. After the flatwork is poured, the contractor blocks or protects the driveway. No vehicle parking is permitted!
12. Expansion joint is required every 100 square feet of flatwork/patio/driveway.
13. The maximum slope of the driveway is 10 percent (24 inches in 20 feet). May vary by municipality and must be accessible to the buyer's lowest car. Examine the driveway slope carefully before and after the pour.
14. Concrete was placed and finished with a light broom finish.
15. Washout and overspill were placed in the designated location.
16. All forms and stakes were removed and stacked in designated area upon completion (or removed completely if contractor is turnkey).
17. Signage and SWPPP (stormwater pollution prevention plan) erosion control were replaced, all signage is replaced and clean, and there is no dirt in the street.

STAGE 9: FINAL GRADE STAGE

TRACTOR: FINAL GRADE

Timing: The final grade is performed after the exterior of the home is completely cleaned from any previous construction activities and the exterior topcoat is complete.

Notes:

- Builder supplies the topsoil, loam, or compost soil material, and the tractor contractor provides the equipment and labor to perform the work.
- Consult the plot plan for grading design for lots with unique topography and/or for wooded lots where tree preservation and drainage considerations may dictate grading design.
- On type A lots (falling from the back to the front of the lot), the material should be delivered into the backyard and graded toward and sloping toward the front/curb.

Grading standard: 5 percent slope away from the house to the side yard swales and 2 percent slope inside the swales to the street

Confirm the following:

1. The grading to exposes the slab 8 to 10 inches at the highest point on the slab. This will be the minimum exposure anywhere on the slab.
2. The drainage is clearly flowing away from the house and all concrete patios and walkways.
3. The lot is smooth, and positive drainage is occurring over the entire lot.
4. The graded surface is smooth and even, with no bowls that might hold water. Minimum grade anywhere is 1 percent.
5. Hand-grading along the slab, walks, utility boxes, and in other tight locations as needed was performed to achieve an even and smooth grade.

6. There was no damage to sewer cleanouts or water meter box, which should be level at grade.
7. There was no damage to the home or air conditioning units and its pad.
8. Splash blocks are reinstalled.
9. Flatwork is left clean with no mud.
10. Erosion control is back in place immediately after the final grade is complete, the signage is replaced, and the street is clean.

STAGE 9: FINAL GRADE STAGE

FRONT YARD TREES

Timing: Once the final grade is complete, you can plant the trees.

Tasks: The tree contractor first machine digs a hole in the yard at the designated location(s) and takes that spaded material off site. Contractor returns with the spaded tree to be placed in the hole. Contractor sand-fills in the hole completely around the tree and stakes and ties the trunk on three sides.

Notes:

- Builder and contractor meet and agree on location(s) per community guidelines and per area allowed by the lot shape.
- Locate all utilities before digging in the yard.
- The yard tree specification is 2 to 4 inches caliper tree as measured more than 3 feet off the ground.
- Trees installed in rock or rocky areas need a hole that is two to three times the depth and width of the root ball size.

Confirm the following:

1. The tree is healthy and conforms to the species and caliper size required by the community and subcontract specification.
2. Tree is placed into the spade dug hole with sand to avoid air pockets around the tree and then watered into place, mulched over, and staked on three sides.
3. All grass and weeds from around the base of the tree were removed, and lower limbs were removed to shape the tree. Treat tree-trimming scars sealed as needed.
4. There is no damage to sidewalks or driveway, and all truck ruts are sanded over.

MEP TRIM STAGE

STAGE **10**

- Final grade survey
- Fences and gates
- Landscaping, irrigation, and sod
- Plumbing trim
- Heating, ventilation, and air conditioning (HVAC) trim and system startup
- Electrical trim
- Low-voltage/security trim
- Weatherstrip exterior doors
- Mirrors and glass doors

The mechanical, electrical, and plumbing (MEP) trim stage installs the finishing touches both inside and outside. The final grade certification will allow for the fence, follower beds, and sod to be installed. The MEP trim follows the usual plumbing-HVAC-electrical progression to complete the systems installation. Only a few finish items remain, such as the vanity mirrors, shower doors, and installation of other theft-prone items such as fancy front doors.

STAGE 10: MEP TRIM STAGE

FINAL GRADE SURVEY

Timing: Schedule the final grade survey after the fence posts are set and before irrigation is installed.

Tasks:

- The survey contractor surveys the grade of the lot to certify that the slope on the lot is 1 percent or greater.
- Confirm that the fence will be located on the property line.

Notes:

- A slope less than 1 percent will fail, and the lot will have to be regraded and resurveyed until the home receives a passing grade of more than 1 percent.
- The final grade survey is retained in the home's permanent files.

Confirm the following:

1. All areas of the lot slope at or greater than 1 percent.
2. The slab elevations at all four corners of the home are noted on the survey.
3. The fence posts are located directly on the property line.

STAGE 10: MEP TRIM STAGE

FENCES AND GATES

Timing: When the exterior topcoat and final grade, and ideally final grade survey, are complete, the house is ready for the fence installation.

Tasks: Contractor locates the property pins and places the fence along the property lines.

Notes:

- Contractor supplies all material and labor required.
- Builder and contractor must confirm that layout conforms to the plot plan and all specialty lot fence requirements are applied.
- The fence should not extend beyond the electric meter to allow access to the meter from outside the yard.
- Wooden fences are "good neighbor" fences styled with solid side of pickets facing opposite directions every other post.

Fence standards:

- Posts are 4 inches by 4 inches and set 24 inches deep in concrete and 8 feet on center.
- Fence has three rails of treated pine.
- Cedar fence pickets are either 1X4 or 1X6.
- Fence is nailed twice on each of the three rails with American hot-dipped galvanized nails **only**.

Confirm the following:

1. Fence was installed plumb and in a straight line.
2. Pickets are in good condition and are twice nailed per rail (six nails per picket, 2 per rail) with approved galvanized nails.
3. Pickets to be installed 2 inches off the ground to allow for sod.
4. Gate(s) were installed per the plot plan, 3 feet wide, level, and close properly.
5. Confirm accessibility to and around the air conditioning (AC) unit(s); add a gate if needed.

STAGE 10: MEP TRIM STAGE

LANDSCAPING, IRRIGATION, AND SOD

Timing: The house is ready for landscaping and irrigation when the final grade survey is complete and passed and before sod.

Tasks:

- The landscape contractor fine grades, prepares beds, and installs the community-specific landscape package, irrigation system, and trees per the subcontract agreement.
- Contactor supplies all irrigation equipment, topsoil, bedding soil and bank sand, sod, plants, and mulch.

Notes: Irrigation may be a community-specific standard feature or an option. A specific number of zones are already included in the subcontract based on community lot size. Oversized lots may require additional irrigation zones to accommodate yard areas per the subcontract agreement.

Confirm the following for irrigation:

1. The installed package meets plant species, size, and count in the subcontract and all applicable community requirements for corner, greenbelt, lakefront, or other amenity lot conditions.
2. Irrigation plan matches the community-specific subcontract in number of zones and other zones added and budgeted during the starts process.
3. Spray or drip system may be used in beds as required by local municipalities. No irrigation spray heads are spraying on the house, sidewalks, or driveway.
4. Community-specific zones were installed and are clearly named and marked on the irrigation control box.

Confirm the following for beds:

Note: *The landscape contractor supplies all topsoil, sand, plants, and mulch.*

1. All bed layouts and sizes must conform to the plot plan, any special requirements for amenity lots or corner lots, and the budget written during the starts process.
2. Bedding soil and sand is well mixed in bed areas. Poorly draining must be properly blended with topsoil and sand, or the plants will die.
3. The plant count, size, and species of plants installed match the community-specific landscaping subcontract.
4. Larger-size plants are staked and tied as needed.
5. Mulch is installed a minimum of 2 inches thick in beds and around new trees.
6. Install downspout extension hoses out of beds with splash blocks as needed to avoid bed washout.
7. All plants were thoroughly watered in after installation, and all flatwork was left clean and free of mud.

Confirm for sod:

Note: *Sod is installed after final grade, fences, landscaping, and irrigation.*

- Sod installation guidelines:
  - » Sod was installed to provide a minimum of 6 inches of exposed slab on all sides.
  - » Sod is flush with flatwork to allow drainage over walks.
  - » Sod is lower than the front fence to allow drainage under fence (fence must not act as a dam).

1. Sod arrives green and healthy-looking unless in the dormant season.
2. Sod is rolled flat over the entire lot. Sand in any low areas prior to installation to ensure a flat lot and lawn.
3. Sod is completely watered in fully after installation. Builder confirms good drainage and no ponding water after sod installation.

4. Builder evaluates weeds in sod; if they are excessive, contractor should weed and feed as needed.
5. Contractor removed all pallets from the site (not part of our trash).
6. All walks are clean, and signage is clean and in place.
7. Sod installation follows the sod installation guidelines.

STAGE 10: MEP TRIM STAGE

PLUMBING TRIM

Timing: Plumbing trim starts after all countertops and backsplashes are complete and tub/shower surrounds are complete.

Tasks: The plumber installs the commodes, tubs and lavatory fixtures, garage disposal, ice maker outlet, kitchen fixtures, dishwasher, and water heater. The drain lines for water heater and furnaces per plan are installed at this time.

Note: Plumbing contractor furnishes all material and labor.

Confirm the following:

1. There are no leaks at any lavatory, kitchen sink, dishwasher, or showerheads.

Note: *Look inside the cabinets for evidence of leaks on the cabinet floor.*

2. Commodes are level and sit stable to the ground. Confirm the good operation and condition and secure installation of all commode seats.
3. All sink water stoppers are functioning properly; adjust as needed.
4. Confirm smooth operation and function of all water control valves and proper siding and mixing of hot/cold.
5. Kitchen spray arm is functioning and retracting completely and properly.
6. Listen for noisy pipes when operating valves.
7. There is adequate pressure at all valves.
8. All fixtures are sealed with plumber's putty to countertops to prevent water intrusion into your cabinets.
9. The gas key for the fireplace is in the wall or secured to the designated location.

STAGE 10: MEP TRIM STAGE

HVAC TRIM

Timing: Once the plumbing trim is complete, the house is ready for HVAC trim. To complete the system startup, the electrical trim will also need to be complete.

Tasks:

- The contractor installs all supply registers, air return grills, and jump duct grills, dryer vent escutcheon, air handler system filters, and thermostat(s). The contractor starts and runs the system to confirm it is functioning to performance standard.
- Contractor supplies or pours the outside pad for the AC, sets the compressor, ties in all necessary copper lines, starts up the system, and tests for proper operation.

Notes: All material and labor is supplied by the contractor.

Performance standard:

- Run the system (heat and cool) for several hours confirm the system meets standard: The temperature in all rooms is within 3 degrees of the thermostat set temperature, as measured 5 feet off the ground in the center of each room.
- Confirm secondary drain line location (side or back over a window) with the contractor prior to starting the work.

Confirm the following:

1. The community-specified thermostat was installed and fits flush to the wall.
2. The community-specified air filter (one-year media filter or monthly filter) is installed standard, as an option. Return air grills open easily and properly.
3. The primary AC drain was correctly installed and secured to a plumbing drain with no crimps in the line, and the secondary drain is in the agreed-upon location.

4. All vents and grills were installed flush and tight on all four sides to the wall or ceiling.
5. The compressor pad was built as agreed, and the compressor is level and solidly secured to the pad.
6. Confirm all refrigerant lines are insulated with 3/8-inch foam rubber all the way to the coil.
7. Check the unit for damage to the grills. All HVAC system components must match the subcontract brand and Seasonal Energy Efficiency Ratio (SEER) rating.
8. System startup is successful, defined as all rooms cooling/heating to within plus or minus 3 degrees of the thermostat set temperature as measured 5 feet off the floor in the center of each room.
9. All trash is in the appropriate receptacle, and all work areas are clean.

STAGE 10: MEP TRIM STAGE

ELECTRICAL TRIM

Timing: The house is ready for electrical trim after plumbing trim is complete and in coordination with the HVAC trim and the tile, paint, and tops are complete. Exterior lights are installed after the masonry or stucco is complete.

Tasks: Contractor trims out all electrical switches, plugs, and power to the appliances, hangs all light fixtures, and makes the electrical connections to the furnace in the attic and the outside AC units.

Notes:

- Contractor supplies all materials and labor for this phase. Depending on the community-specific electric subcontract, contractor may or may not be the light fixture supplier.
- Electrical contractor needs to connect the HVAC for the system startup normally by making a second trip to the site during the MEP trim. While there is a second trip, there is only one electrical trim payment.

Confirm the following:

1. All light fixtures and fan switches work properly. Confirm logic on three-way switches and that all fan speed controls are functioning.
2. Hot punch all electrical outlets. Exterior ground-fault circuit interrupter (GFCI)-protected plugs are working properly, and all exterior plugs have their cover that locks shut.
3. All light fixtures, plugs, and switch plates are plumb with no missing screws, and all fit flush to the wall or ceiling.
4. All light fixtures were installed complete with all their parts and not missing parts/glass, and fixtures match the specifications.
5. All recessed cans fit snug to the ceiling, plumb and flush.

6. The smoke detector, doorbell, and attic lighting are functioning properly.
7. All installed appliances are connected and working.
8. The connection to the AC units located outside is complete.
9. The electrical panel is labeled clearly and legibly, identifying all locations per switch.
10. All boxes, wrapping, packaging trash, and debris was removed from the job site by truck, and all work areas are clean.

STAGE 10: MEP TRIM STAGE

LOW-VOLTAGE TRIM

Timing: Low-voltage security and structured wiring may be trimmed at the same time as the electrical trim.

Tasks: The prewired security system opening contacts and control touch pads, audio speakers, structured wiring, and other audio-visual features of the home are trimmed out at this time.

Notes: In many communities, the electrical contractor and low-voltage contractor are the same contractor, and in that case, the low voltage trim occurs simultaneously with the electrical trim.

Confirm the following:

1. All community standards and buyer options are included and trimmed.
2. The system is functioning as designed.

STAGE 10: MEP TRIM STAGE

WEATHERSTRIP EXTERIOR DOORS

Timing: The home is ready for weatherstripping the exterior doors at the same time or prior to the HVAC trim.

Tasks:

- Weatherstrip all doors opening to a non-AC location, including the attic walk-in door or pull-down attic stairs, the garage man-way door, or any other door from the conditioned space inside the home to a non-air-conditioned space.
- Contractor installs the cushion bronze strips on the sides and top of the jamb, aluminum threshold with a matching flange at the bottom of the door.

Notes: Some exterior doors arrive with weatherstripping factory installed, but for those that do not, the weatherstripping is installed in the field at this time.

Confirm the following:

1. The doors operate smoothly and close flush to the jamb. There is no daylight visible around any side or top of any door.
2. Weatherproof adhesive is used beneath the threshold/door pans, and the threshold is also caulked at the sides to the door jambs. All bottoms of wooden doors are sealed to prevent wicking.
3. All weatherstripping is well fitted to the door and jamb on all three sides.
4. The attic walk-in and/or pull-down access door is weatherstripped.
5. All work areas are clean, and debris was placed in the designated receptacle.
6. If the water heater is in a closet, the door is also weatherstripped.

STAGE 10: MEP TRIM STAGE

MIRRORS AND SHOWER DOORS

Timing: When the plumbing trim is complete, the house is ready for bath mirrors and shower doors and enclosures.

Tasks: The contractor field measures all parts for the bath mirrors and glass and fabricates and installs each according to the plans and subcontract agreement.

Notes:

- Glass contractor provides all materials and labor for this work.
- Confirm at the time of the field measurement:
    - » All buyer's glass options
    - » Mirror heights and edge styles,
    - » Shower door hinge side
    - » Shower door style and color
    - » Light fixture, plug, or switch locations that may conflict with mirror or glass locations

Confirm the following:

1. Shower enclosure matches the community color and design specification, or the optional frameless style as indicated on the buyer selection summary.
2. Shower door sweeps evenly and smoothly over the tile surface to prevent water leaking from under the door.
3. All shower door frame mitered joints are smooth and not hanging and fully sealed with siliconized clear caulk.
4. Mirror and door installations are plumb and level to the eye.
5. Mirrors connected with 25-pound clips attached to studs. Never use mastic or other adhesives in place of clips.
6. There is no scratched glass or mirrors.

INSPECTION-READY STAGE

- Window inspection and screen install
- Install wooden front door (option/community-specific)
- Appliance installation
- Finished floors
  - Wood floors
  - Carpet
- Garage door opener(s)
- Paint final and paint touch-up
- Final exterior and garage clean
- Final interior and window clean

The inspection-ready stage is the final stage of installing the final finishes of the home that are most sensitive to damage, such as finished flooring, garage door openers, and appliances. The last and final stage involves a lot of punchout, paint touch-up, and interior and exterior final cleans.

STAGE 11: INSPECTION-READY STAGE

WINDOW INSPECTION AND SCREEN INSTALL

Timing: The home is ready for the window inspection and screens as one of the very final steps in completing the home.

Tasks: The window supplier inspects the operation and locking of every window on the home, installs the screens, and ensures that all windows are in perfect working and cosmetic condition.

Note: Walk with window company representative upon arrival to inspect the windows and point out any known window issues that must be addressed. Show the problem to the representative (not just tell them about it) to ensure a clear understanding and participate in the solution decision.

Confirm the following:

1. Screens are installed on all operable windows.
2. All windows are operating per design standards; repair as needed.
3. Inspect and confirm all previously discussed issue windows and plan additional repairs as needed are complete.
4. All windows are properly lubricated as part of the inspection.
5. There is no cracked or broken window glass.

STAGE 11: INSPECTION-READY STAGE

INSTALL WOODEN FRONT DOOR

Timing: In communities where the stained mahogany door with a leaded-glass-insert front door is standard or where a buyer has selected this option. These doors should be installed only immediately prior to final inspection of the home.

Tasks:

- The contractor provides the door and all materials and labor for the door's installation and weatherstripping.
- At the time of the release, the builder confirms hardware selection, door style, stain color, and finish as stated on the selections summary.

Note: Decorative wooden entry doors may be a standard specification in some communities and an option in other communities. These doors can take up to two weeks to fabricate; call early and confirm style and color selection and swing at the time of the release.

Confirm the following:

1. Prior to installation, the door jamb is plumb, square, and level.
2. Prior to installation, the door matches the selections summary.
3. Before the contractor leaves the site, inspect the door for proper fit and operation and confirm proper key fit and locking of dead bolt and no visible light from around any side of the door.
4. The bottom of the wooden door is properly sealed. Wicking from the bottom can otherwise be a warranty issue.
5. Contractor removes the "construction door" from the job site.

STAGE 11: INSPECTION-READY STAGE

APPLIANCE INSTALL

Timing: Appliances should be installed only if the home is sold and the buyer walks and closing are imminent.

Tasks: The appliance installation company delivers the standard kitchen appliances or other optionally selected appliances and installs and test operates each to confirm proper function at the time of the install.

Notes:

- The appliances are installed by an independent appliance contractor hired by the appliance vendor selling the builder the appliances.
- The contractor provides all materials and labor for the installation.

Confirm the following:

1. All appliances match buyer selections and options per the selections summary.
2. All appliances are in perfect condition, free of scratches, after the contractor has completed the installation.
3. The microwave and oven are installed level and centered on each other in the tower.
4. All appliances are tested and working properly, and the vent hood is properly tied into the venting.
5. All boxes and packaging were removed from the home, and the work areas are clean.
6. Leave all appliance manuals and warranty documents in a designated single kitchen drawer.

STAGE 11: INSPECTION-READY STAGE

FINISHED FLOORS: WOOD FLOORS

Timing: Wood floors are one of the very final tasks in completing the home and are installed only when the home is completely punched out and no more repairs or installations are anticipated.

Tasks:

- Builder and contractor collaborate to confirm that the moisture content of the concrete or wooden subfloor are less than 4.5 percent moisture content in the areas to receive wood flooring and that the flooring material has been climatized to the house according to the manufacturer's recommendation.
- Contractor floats the floor as needed to achieve a level floor. When the floor is ready, the contractor installs the wood floors in the designated areas according to the manufacturer's installation recommendations.

Notes: The home and flooring must be in a steady state regarding moisture. The home's air conditioning (AC) system must be running a minimum of two days prior to installation, and if possible, the flooring will also be in the home in open boxes for two days. Follow all manufacturer's recommendations exactly.

Confirm the following:

1. Prior to installation, the flooring matches all selections and locations designated in the selections summary.
2. Floors are floated level as needed to ensure level floor prior to installation. Floors are level within 1/8 inch over any 10-foot span or radius.
3. Contractor uses only manufacturer's recommended adhesive applied using the manufacturer's recommended trowel size.
4. Seams are tight, and there are no defective boards.
5. Home is left clean, and the floors are completely protected by paper cover.

6. All unused flooring boards, at least 50 square feet or more, are stored in attic for future use by the homeowner or other warranty work that might be required.
7. Contractor installed the shoe molding and nail joints and puttied all raw ends.
8. No glue is visible on the baseboards, and all surrounding areas were left clean.

STAGE 11: INSPECTION-READY STAGE

FINISHED FLOORS: CARPET

Timing: The carpet is one of the last things to go into the home, installed only when the home is completely punched out and no more repairs or installations are anticipated.

Tasks: Contractor scrapes and sweeps the installation areas. Tack strips are placed no more than 1 inch from the walls. Padding is installed using glue or tape. Carpet is stretched into place and secured onto the tack strips.

Notes:

- The contractor supplies all material and labor for carpet.
- Verify the carpet and pad and locations designated match those identified on the selections summary.

Confirm the following:

1. Prior to carpet installation, all divots or other inconsistencies in the concrete or subfloor are located and floated smooth.
2. Transition lines between hard surfaces and carpeting are built up as necessary by using float or double-tack strip.
3. Carpet is secured around the entire perimeter with no protruding nails or tact strip edges.
4. Dye lot colors are within tolerance; inspect as the carpet is rolled out.
5. Carpet nap is the same on both sides of and at the seams.
6. All seams are tight and secured completely and are hardly visible.
7. Stair treads are wrapped completely and tightly with no visible nails.
8. All doors that were removed have been rehung.
9. Place usable-sized scraps in the attic for the homeowner.
10. Carpet is vacuumed, and all small scraps were removed from the home.

STAGE 11: INSPECTION-READY STAGE

GARAGE DOOR OPENER(S)

Timing: The garage door opener(s) may be installed any time after all the mechanical trims are complete and the permanent power is on the house but are usually installed only immediately prior to the closing to avoid theft.

Task: Garage door contractor installs the opener and confirms proper operation.

Note: Garage door contractor supplies all materials and labor for this work.

Confirm the following:

1. The doors open and close smoothly using the opener and seals completely and flush to the slab and flush to all three sides of the garage wall.
2. The electronic eyes are adjusted and working properly.
3. The pull cord emergency release operates properly.
4. The two remotes per door and manuals are placed in the designated kitchen drawer with other manuals and homeowner materials.
5. Work area was left clean, and all debris is in the appropriate receptacle.

STAGE 11: INSPECTION-READY STAGE

PAINT: PAINT FINAL

Timing: When all work is complete on the home and there are no more repairs or work to be performed by any other contractors

Tasks:

- Paint contractor reviews and inspects all painted walls and trim for any imperfections or any incomplete painting work and touches up or repaints as needed to satisfy the builder and buyer.
- This task will likely require multiple trips after each inspection or walk and after any required repair is complete.

Note: In some cases, the paint contractor may attend certain buyer walks to immediately paint any discovered blemishes during the walk and to the buyer's or inspector's satisfaction.

STAGE 11: INSPECTION-READY STAGE

FINAL EXTERIOR AND GARAGE CLEAN

Timing: The final exterior clean should be scheduled as soon as you are confident that there is no more exterior work to be done.

Tasks:

- Contractor cleans all flatwork of any mortar, paint dropping, glue, dirt, and tire marks using a power washer and hand scraper, brush, paint thinner, etc., as needed. Lightly power-wash the garage door and any brick areas as needed.
- Contractor inspects the roof and exterior cladding of the home for any overspray or other areas of the exterior that need to be cleaned off brick or stone surfaces.
- Contractor inspects the flower bed and sod areas and ensures they are fresh and clean-looking, with no sunken areas.
- Contractor cleans the entire lot, removing any visible remaining construction trash, packaging, and micro-trash on the lot or any neighboring lot.

Note: This clean is in preparation for the buyer walks and is the last time to inspect and deliver the near-perfect home.

Confirm the following:

1. The roof is cleaned and looks perfect from the street. Replace any damaged shingles if needed.
2. No micro-trash is visible in the beds and lawn, and there is no blown trash along the fence in the yard or in neighboring yards.
3. There is no mud or dirt on sidewalks, meter boxes, or nearby utility boxes.

4. Sunken or low areas on the lot, if any, are filled with sand as needed so that the sod surface is smooth to the eye.
5. Bed mulch is "fluffed up" to appear freshly mulched, with no footprints in the bed's mulch.
6. Garage, porch, and patio are completely clean and power-washed as needed. Clean all spider webs from the ceiling corners, and confirm walls and ceiling of the garage is clean like new.

STAGE 11: INSPECTION-READY STAGE

FINAL INTERIOR AND WINDOW CLEAN

Timing: The final interior clean may begin when all interior and construction punch and touchup work is complete. This is the very last phase of construction.

Tasks: The cleaning contractor cleans all windows inside and out, sweeps and vacuums the entire home, mops all hard-surface floors, cleans the kitchen and all bathrooms completely, and cleans out the return air ducts.

Note: Ideally, the clean contractor is the last subcontractor in the home.

Confirm the following:

1. All window track, frames, and glass are clean and not scratched.

   Note: *Scratching glass is a very expensive error; only experienced window cleaners can use razors on glass.*

2. There are no signs of mud or marks on walls, doors or casing, windowsills, cabinets, or trim woodwork, including tops of base boards. All woodwork is clean.
3. All mirrors are clean, with no scratches or streaks.
4. All stickers were removed from light fixtures, cabinets, and hardware, and there are no visible manufacturer's stickers of any kind.
5. Cabinets are clean inside and out.
6. The water heater and air handler in the attic are wiped clean. You will be demonstrating them to the buyer, and they need to look brand new.
7. The attic is swept, and the walkway is clear.

SIGNOFFS AND BUYER WALKS STAGE

STAGE 12

- Internal management walk
- Final inspections
    - » Independent code inspector's (ICI) final quality walk and signoff
    - » Buyer's independent inspector (if there is one)
    - » City occupancy/final inspections
    - » Energy inspection and program audit/certifications and infrared scan envelop inspection (as required locally)
- Buyer demonstration walk
- Buyer final walk

The entirety of the construction process has been leading up to this home delivery stage. The builder completes all self-inspections before undergoing a series of manager and independent code and municipal inspections. Each of those inspections is leading up to the buyer walks, the success of which will rely on the builder's interpersonal and organizational skills as much as or more than the builder's technical skills. The home delivery process is part of what makes homebuilding a retail business, and builders who delight in delighting clients will love this final stage of the construction process.

The end game of the buyer walks is to gain the buyer's signature, indicating acceptance of the home as complete and thus turning the buyer into a homeowner.

STAGE 12: SIGNOFFS AND BUYER WALKS STAGE

MANAGEMENT WALK

Timing: The management walk is an internal walk and the last set of company eyes on the home at completion. The walk may be performed by the project manager, the warranty manager, or another designated quality compliance manager. The manager inspects the home using the final inspection checklist.

Goal: This walk is to confirm that the home is completely ready for various third-party inspection(s) and the buyer walks.

Notes: When the manager signs off on the home, the builder may call for the independent code inspector's inspection and proceed with all other required third-party inspection(s).

Confirm the following:

1. Have the following items on hand:
    - » Buyer selection summary sheet
    - » Plot plan and a full set of plans
    - » All inspection pass letters or green tags
    - » Any house certificates: cable stress, insulation, etc.
    - » Any third-party inspection reports previously produced
    - » Buyer correspondence indicating their special concerns
    - » All remote controls for fans, garage door openers, fireplace controller, etc.
    - » All manufacturer's operation or warranty manuals
2. The manager uses the inspection checklist, but moreover, the home is walked the way the buyer will walk and inspect the home: visually.
3. Builder promptly corrects any remaining items of concern discovered and listed from the management walk.

STAGE 12: SIGNOFFS AND BUYER WALKS STAGE

INDEPENDENT CODE INSPECTOR: FINAL QUALITY WALK

Timing: Schedule the ICI final inspection immediately upon completion of any items that came up on the management walk. Schedule as early as possible to complete items from the ICI final prior to the municipal inspections.

Tasks:

- The ICI final walk covers the 500-plus items on the company's preset ICI final walk list.
- Builder promptly corrects any defective items from the ICI inspection list.

Note: The ICI inspects the home technically and relative to current code and creates a code compliance report that will become part of the home's permanent records.

Confirm the following:

1. Have the following items on hand:
    - » Buyer selection summary sheet
    - » Plot plan and a full set of plans
    - » All inspection pass letters or green tags
    - » Any house certificates: cable stress, insulation, etc.
    - » Any third-party inspection reports previously produced
    - » Buyer correspondence indicating their special concerns
    - » All remote controls for fans, garage door openers, fireplace controller, alarm fob, etc.
    - » All manufacturer's operation or warranty manuals

STAGE 12: SIGNOFFS AND BUYER WALKS STAGE

CITY OCCUPANCY/FINAL INSPECTION

Timing: Once the house has passed the management walk and the ICI inspection, the house is ready for the city's final inspection.

Notes:

- If the home is sold, schedule the buyer demonstration walk at this time in anticipation of passing all city inspections.
- Always meet the inspector at the home and accompany the inspector on the inspection.

Confirm the following:

1. Have the following items on hand:
   - » Buyer selection summary sheet
   - » Plot plan and a full set of plans
   - » All inspection pass letters or green tags
   - » Any house certificates: cable stress, insulation, etc.
   - » ICI code compliance report
   - » Any third-party inspection reports previously produced
   - » Buyer correspondence indicating their special concerns
   - » All remote controls for fans, garage door openers, fireplace controller, etc.
   - » All manufacturer's operation or warranty manuals
2. Property identification and required postings are in place and visible. Address is clearly visible from the street.
3. The exterior front yard and backyard, the garage, and everything about the home should be swept and clean, with no debris anywhere, and the home is ready for any inspection.

STAGE 12: SIGNOFFS AND BUYER WALKS STAGE

BUYER FIRST WALK: DEMONSTRATION WALK

The buyer first walk, also known as the *buyer demonstration walk,* has several important goals that must be accomplished to deliver the home to the new owner.

Builder demonstrates the home to the buyer following and covering the operation of all mechanical devices in the home and demonstrating the complete, good, and ready condition of the home.

The buyer demonstration walk is the buyer's one last opportunity to inspect the home's cosmetic condition to be corrected by the builder.

Key goals:

1. **Demonstrate and train the *function and operation of the home itself.*** Covered items include mechanical, electrical, plumbing, irrigation, yard lights, and audiovisual options.
2. **Demonstrate and train the *manufacturer's recommended and otherwise necessary maintenance and care for each component of the home.***
3. ***Document any cosmetic or functional items identified by the buyer at the demonstration walk.*** All identified items are documented on the buyer's acceptance form and must be addressed by the builder during the next week and prior to the final acceptance walk.

> **Note:** This is the one and last time that the buyer may inspect for general cosmetics of the home. This gives the builder clear sight on exactly what needs to be completed to satisfy the buyer.

4. ***Review the going-forward warranty process.***

   This is a new phase in the builder/buyer relationship, and the buyer needs to understand how to get service in the future and that the builder is no longer the point of contact with the company.

5. ***Earn the buyer's trust and confidence in their home and in you as their builder.*** A goal of the first walk is to give the buyer peace of mind.

Builder is responsible for scheduling and conducting the demonstration walk.

1. **Schedule the walk time with the buyer at least seven days prior to the day you want the walk to occur and not fewer than 14 days before the anticipated closing date.** This gives the builder time to address all buyer items properly and have zero items remaining at the final walk. Ideally, set appointments for both the first walk and the final walk at this time. Allow time and be flexible to have a smooth and unrushed meeting.
2. **Arrive at the home early to the meeting, turn on all lights, and confirm everything is perfectly clean.** Look in every room and closet and the garage. Walk the sidewalks and confirm everything is still perfectly clean.
3. **Have a written agenda for the walk and follow it completely** to ensure the entire home and all its components are covered with the buyer. Review the agenda with the buyer to start the meeting. Use the agenda/checklist throughout the meeting to ensure the walk's efficiency and effectiveness in what can be a multihour home demonstration and quality review.
4. **Builder controls the mood, pace, and flow of the walk.** The walk is more about discovering the buyer's wonderful home than it is about finding hidden problems. The builder's people skills are key to a successful walk.
5. **Upon arrival, meet the buyers in the driveway or at the entry.** Lead the group to the kitchen area and start the conversation about the agenda and goals of today's walk. In the kitchen, explain the walk process and all the paperwork associated with today's walk that will need to be signed at the end of the walk.

6. After the discussion in the kitchen, the builder will direct the buyer back outside and **start the walk outside the home at the curb and walk and demonstrate the complete exterior first,** then proceed inside to do the same.
7. **The demonstration of the home and its components is an active demonstration.** Physically open and close or turn on and off whatever you are demonstrating. **Make the buyers open and close and operate whatever is being demonstrated. Do not simply explain how things work. Show them.**
8. **At the conclusion of the walk,** the group should reconvene in the kitchen and agree on the final list of items to be addressed by the builder in the next seven days. **Buyer and builder both sign and date the appropriate forms.**
9. **Reschedule or reconfirm the appointment for the final acceptance walk,** ideally the day before the closing. Review the goal and limits of the final walk with the buyer. The goal and limit of the final walk is only to review the items listed from today's walk, not make a new list.

STAGE 12: SIGNOFFS AND BUYER WALKS STAGE

BUYER FINAL ACCEPTANCE WALK

Timing:

- The home is ready for the buyer's final acceptance walk as soon as all items identified and listed on the buyer's demonstration walk list have been resolved, repaired, or replaced and the home is recleaned.
- Schedule the final acceptance walk to allow as much time as possible. Buyer must never feel rushed or pressured for time during the final acceptance walk.

Tasks:

- Re-walk **only** those items from the buyer's first walk.
- Builder must control the meeting and ensure that no second list is started at this walk.

Goal: The goal is to close every home with zero outstanding items. This means that at the final acceptance walk, the homeowners agree, confirm, and sign off that the home is 100 percent complete; accept the quality, workmanship, and cosmetics of the home; and have no outstanding issues with the builder, the home's materials, or workmanship and that all included features and selected options are installed to their satisfaction.

At the end of the meeting, the buyer must sign the certificate of completion indicating their acceptance of the home. If there are unresolved items remaining at the end of this walk, the buyer must accept everything else on the home, save and except those identified items.

Signing the certificate of completion is today's key goal.

Notes:

1. **Builder controls the focus of the final acceptance walk to ensure that only those outstanding items identified at the first walk's acceptance form are reinspected on this final walk.** This point should be clear from communication at the demonstration walk; blending this into your message early and often is important.
2. For **known problem buyers,** it is advised that the warranty or project manager attend this walk to ensure that warranty coverages are clear on items of concern. Manager should document and place notes in the house file any discussions or agreements with the homeowner from this walk.
3. **If there remain any buyer items after the final walk:**
    - The home may still close with incomplete item(s) from the demonstration walk.
    - If builder thinks the buyer may never sign after closing, contact your manger immediately. The closing could be delayed to complete items if needed or clarify everyone's expectations regarding the repair.
4. **If the home closes with items, it remains the builder's responsibility to complete the outstanding items and secure the buyer's signature on the certificate of completion form.**

Construction Glossary

A

Air conditioner (AC) condensing unit: Consisting of the condensing coil, fan, and compressor.

AC evaporator: This coil is located in the attic in the air handler/furnace air stream.

AC disconnect: Code requires that all AC condensing units be in the line of sight of the breaker panel unless they are equipped with a dedicated 230-volt disconnect.

AC primary drain: The main drain for the AC, which extends from the evaporator coil to a sewer line; usually tied in at a bath sink.

AC secondary drain: The pan overflow that is terminated over a window as a backup drain.

Adhesive: Any substance that holds the surface of two materials together; common adhesives in residential construction are cements, glues, and mastics.

Aerators: A device on water faucets designed to screen foreign matter and prevent splashing.

Airway: A space between roof insulation and roof boards for movement of air.

Ampere (amp): The amount of electricity that flows through a conductor or circuit; watts divided by volts equal amps.

Anchorage: A device used to anchor the post-tensioning steel (tendon) to the concrete member.

Anchor bolt: A bolt that fastens base plates to concrete; usually a J-bolt; must be 0.5 inches in diameter and embedded in concrete by seven inches.

Anchor; dead end: The device at the non-moving end of a posttension cable, installed at the factory, by which the cable is secured to the inside of the forms.

Anchor; live end: The device put on the live or moving end of the posttension cable in the field, prior to installing the pocket former and by which the live end is secured to the forms.

Apron: The flat member of the inside trim of a window placed against the wall immediately beneath the stool.

Art niche: An inset designed into a Sheetrock wall designed as a place to hang art.

Astragal: A molding attached to one of a pair of swinging doors, against which the other door strikes; may be referred to as *T-astragal* because of its shape.

B

Backfill: The replacement of excavated earth into a pit or trench or against a structure.

Back splash: A short extension of the countertop up the wall to the wall cabinet.

Baluster: A small pillar or column used to support a rail.

Barge board: The board at the gable just under the edge of the roof behind the fascia, which supports the overhang.

Base board: The molding covering the joint between a finished wall and a floor.

Base shoe: Molding used next to the floor on interior baseboard; usually seen only where hard surface floors exist.

Batter board: A temporary framework used to assist in locating the corners when laying out and squaring a foundation formwork.

Bay window: Any window space projecting outward from the walls of a building, either square or polygonal in plan.

Beam: One of the principle horizontal supporting members in a building; generally, in residential construction, beams are of the built-up or laminated beam variety; the built-up beam is usually two 2x members with either a steel or plywood flitch plate bolted or nailed between them, done in the field; more commonly used is the laminated beam, sometimes known as the *glue-lam* or *versa-lam beam*, depending on the manufacturer.

Beams, foundation: Trenches dug and reinforced prior to pouring the slab on grade to provide additional support and bearing, lending structural strength around the perimeter.

Bearing wall or partition: A wall that supports any vertical load in addition to its own weight.

Bi-fold door: An interior door unit that is hinged in the middle and folds against the jamb.

Benched lot: A lot set flat and to the engineer's specified elevations; ready to build without further excavation.

Blind nailing: Nailing in such a way that the nail heads are not visible on the face of the work.

Board foot: The equivalent of a board 1 foot square and 1 inch thick, or 144 cubic inches.

Bow: Warp upward along the length of a piece of lumber laid flat.

Braces: Pieces fitted and firmly fastened to two others at any angle in order to strengthen the angle.

Bracket: A projecting support for a shelf or another structure.

Breaker: A safety device located in the breaker box that is designed to trip if a short circuit condition or an overload exists on any one electrical circuit.

Breaker box: The metal box, sometimes referred to as the *electrical panel*, where the service entrance terminates and is subdivided into individual circuits, either 115 or 230 volts, servicing the entire structure; each individual breaker must be labeled with the portion of the home served.

Break joints: To arrange joints so that they do not come directly under or over the joints of adjoining pieces, as in shingling, siding, etc.

Brick frieze: Horizontal piece of cornice trim located at the top of an exterior wall where the brick meets the soffit of the home; usually Hardie plank that is 1 inch by 4 inches per series specifications.

Brick ledge: The 1.5-inch-by-5.5-inch drop on the edge of the foundation designed to receive brick; created by the use of a 2-inch-by-6-inch nailed flat against the top of the forms when forms are floated before the pour.

Brick tie: A corrugated, galvanized metal strip that is nailed to the frame and bent out to be embedded in the mortar joint of a brick veneer wall to hold the brick securely to the structure; usually set on a 24-inch grid.

Brick veneer: A facing of brick laid against a frame wall construction that has no structural integrity with regard to load-bearing properties; held in place by brick ties secured to frame.

British thermal unit (BTU): The amount of heat that is required to raise the temperature of one pound of water one degree Fahrenheit at a specific temperature.

Builder's level: Also known as a *sight level*; a tripod-mounted instrument used to shoot grades through the use of a calibrated stick as well as angles on the ground; unlike a transit level, however, it operates only in the horizontal plane, relying on the pole to establish the grades.

Building code: A collection of legal requirements the purpose of which is to protect the safety, health, and general welfare of those in and about buildings; we currently use the International Residential Code for One and Two-Family Dwellings 2000, which is commonly referred to as the *IRC*.

Bull float: A large float tool on a long handle used for rough finish of concrete to a level condition in preparation for curing and troweling; follows the use of a screed.

Bull rock: The 1.5-inch-by-3-inch multishade brown decorative landscape stone.

C

Cable, posttension: Manufactured seven wired 270K strand, greased and coated with plastic sheathing; a 0.5-inch cable has a breaking strength of approximately 41,000 pounds.

Cantilever: Any part of the structure that projects beyond its main support and extends without a bearing wall on its projecting end.

Cased opening: An interior doorway or window not furnished with a door; generally cased out with a non-stopped jamb and door casing.

Casement or casement window: A window in which the sash opens upon hinges.

Cat head: The aluminum or cast-iron clamping device applied outside the form, on the post-tension cable's live end, to maintain tension, or "stretch," on extended cables.

Ceiling: The finished lining of the surfaces that enclose the upper side of a room; typically drywall with a stamped finish.

Chair: Supports used to lift the post-tensioning cable above the soil surface to achieve good concrete coverage and maintain proper position of the cable in the center of the slab.

Chamfer: A beveled surface cut upon the corner of a piece of wood.

Checks: Splits or cracks in a board, ordinarily caused by seasoning.

Cleat stock: Wooden shelf support.

Closing: The date of ownership of the home after its completion; the builder and client exchange keys and money at the closing, which usually occurs at a title company.

Collar beam: A tie beam connecting the rafters considerably above the wall plate that is intended to prevent the walls from being pushed outward; also called a *rafter* or *collar tie*.

Columns: A support that is square, rectangular, or cylindrical in section for roofs, ceilings, etc.

Common rafter: Any rafter that connects at one end to the ridge and the other to the top plate.

Compacted or compactible fill: Soil or dirt used to support the foundation or flatwork.

Concrete: An artificial building material made by mixing Portland cement and sand with gravel, broken stone, or other aggregate and sufficient water to cause the cement to set and bind the entire mass.

Condensation: In a building, beads or drops of water and frequently frost in extremely cold weather that accumulate on the inside of the exterior covering of a building when warm, moisture-laden air from the interior reaches a point at which the temperature no longer permits the air to sustain the moisture it holds.

Conductors: Individual current carrying or ground wires in a circuit; 115-volt circuits usually consist of three wires.

Corbel out: To build out one or more courses of brick or stone from the face of a wall to form a support for timbers.

Corner bead: A strip of formed galvanized iron or polyvinyl chloride (PVC) placed on corners after drywall is applied to reinforce them.

Corner boards: Used to trim for the external corners of a house or other frame structure.

Corner braces: Diagonal braces let into studs at not greater than a 45-degree angle to reinforce corners of frame structures.

Cornice: Term used to describe all actions of the trade that "dry-in" a home; involves installation of windows, doors, sheathing, siding, exterior finished trim, and louvers; also refers to the molded projection that finishes the top of the wall of a building.

Counterflashing: Strips of metal used to prevent water from entering the top edge of the vertical side of the J flashing; sawn into the brick and sealed and overlap but are not attached to the J flashing to allow expansion and contraction without danger of breaking the flashing.

Countersink: To drive the fastener into a recess below the wood's surface.

Course: A single row of an exterior finish material, such as bricks, shingles, or siding.

Cove molding: A three-sided molding with concave face used wherever small angles are to be covered.

Crawl space: A shallow space below the living quarters of a house; generally not excavated or paved and often enclosed for appearance by skirting or facing material.

Cricket: A small drainage diverting roof structure of single or double slope placed at the junction of larger surfaces that meet at an angle; also called a *saddle*; commonly found behind chimneys.

Cripple stud: Stud that is cut shorter than a precut stud and is used to support areas under and above windows or doors and the sides of any opening to support headers; also called a *jack stud*.

Crown: The high point of a piece of lumber with a bend in it or the apex of the curve of the annular growth rings in a piece of framing material when viewed from the end.

Crown molding: A molding used on cornice or wherever a large angle is to be covered.

Cup: The warping of a piece of lumber that, when laid flat, results in the linear edges being higher than the center.

D

Damper: A metal "door" placed within the fireplace chimney; usually closed when the fireplace is not in use.

Dead bolt: An exterior security lock installed on the exterior entry doors that can be activated only with a key or thumbturn.

Dead end: The end of a posttension cable to which a permanent casting is attached and that is nailed to the inside of the form using 20d nails.

Deadening: Construction intended to prevent the passage of sound.

Dead load: An engineering term used to describe the weight of all structures in place.

Dead wood: Blocks nailed to frames as needed for future nailing/screwing, in the absence of other framing members.

Deck, decked: To install the plywood or water board sheeting on the floor joists, rafters, or trusses.

Dedicated circuit: An electrical circuit that serves only one appliance (i.e., dishwasher) or a series of electric heaters or smoke detectors.

Direct nailing: To nail perpendicular to the initial surface or to the junction of the pieces joined; also termed *face nailing*.

Diverter: Sheet metal bent to a 90-degree angle that is positioned on a roof to divert water away from doorways or other undesirable areas.

Door jamb: The surrounding case into which and out of which a door closes and opens.

Dormer: An opening in a sloping roof, the framing of which projects out to from a vertical wall suitable for windows or other openings.

Double glass: Window or door in which two panes of glass are used with a sealed air space between; also known as *insulating glass*.

Double-hung window: A window with two vertically sliding sashes, both of which can move up and down.

Dry-in: To install the black roofing felt (tar paper) on the roof.

Drywall (or gypsum wallboard [GWB], Sheetrock, or plasterboard): Wall board or gypsum, which is a manufactured panel made out of gypsum plaster and encased in a thin cardboard.

E

Easement: A formal contract that allows a party to use another party's property for a specific purpose; for example, a sewer easement might allow one party (such as the city) to run a sewer line through the property.

Eave: The margin or lower part of a roof projecting over the wall.

Elevation: A specific view of the exterior of a structure (front, right, left, or rear); a specific view of the interior of a structure (kitchen cabinets, bookcases, etc.).

Elongation of cable: The increase in tendon length due to the mechanical stressing of the cable; for a 0.5-inch cable, the elongation is 0.0833 inches per foot of tendon length, or approximately 8 inches in 100 feet.

Escutcheon: Trim shield used around pipes to cover pipe transition to the wall.

Expansion joint: An elastomeric caulk, treated lumber, or bituminous fiber strip used to separate brick or units of concrete to prevent cracking due to expansion.

F

Face nailing: To nail perpendicular to the initial surface, or to the junction of the pieces joined; also termed *direct nailing*.

False rafters: Any 2-inch lumbers used to construct a sloped ceiling.

Fascia: A flat member of a cornice or other finish, generally the horizontal board of the cornice, across the ends of the rafters and to which the gutter is fastened.

Finger joint: A factory method of cutting, fitting, and gluing short lengths of lumber together to make longer, straighter lengths; may be used **only** vertically (in compression) and not horizontally (in tension).

Fire blocking/fire stops: Draft stops made of materials designated in the IRC to serve as impediments to the travel of a possible fire from one floor to the next or from one area to another; the only material we use on the interior of our structures that is fireproof is gypsum board or Sheetrock.

Fire box: The portion of a fireplace that contains the fire; must be UL-rated and UL-approved.

Fire stop: A solid, tight closure of a concealed space; placed to prevent the spread of fire and smoke through a space.

Flagstone (flagging or flags): Flat stones, from 1 to 4 inches thick, used for rustic walks, steps, floors, and other items; usually sold by the ton.

Flashing: The material used and the process of making watertight the roof intersections and other exposed places on the outside of the house.

Flashing J: Galvanized metal flashing used on the roof to channel water away with a J hook bent in the portion under the shingles; used around chimneys and adjacent to walls, dormers, or any time a vertical structure rises and abuts a rake.

Flashing Z: Galvanized flashing used where one piece of 4-foot-by-8-foot siding is applied directly over and abuts another.

Flat paint: An interior paint that contains a high proportion of pigment and dries to a flat or lusterless finish; not a washable or scrubbable product.

Flitch beam: A built-up beam formed by an iron plate or piece of plywood of the engineered thickness being sandwiched between two 2xes; the entire unit must be securely bolted or nailed together as specified.

Flitch plate: The plate sandwiched in a flitched beam.

Float: A term used both in concrete and drywall work; in drywall finishing, the process of applying drywall mud with a wide knife after the tape has been bedded; in concrete, the process of applying the float boards to the form before the pour and the act of rough smoothing the poured concrete.

Flue: The opening in a chimney through which smoke and heat passes; in the case of a prefabricated fireplace, it is a double or triple wall capped with a chimney cap; most of these are located in wooden chimney chases.

Flush: Adjacent surfaces even or in the same plane.

Footing: An enlargement at the lower end of a wall, pier, or column to distribute the load.

Form bumping/vibrating: The act of hitting the forms with a sledgehammer or vibrating the concrete with a mechanical vibrator to move stones in away from the form and release air pockets in an effort to prevent honeycomb effects.

Foundation: The part of a building or wall that supports the superstructure.

Frame: The surrounding or enclosing woodwork of windows, doors, etc., and the timber skeleton of a building.

Framing: The rough lumber structure of a building, including interior and exterior walls, floor, roof, and ceilings.

Framing systems:

- **Platform:** A system of framing a building on which floor joists of each story rest on the top plates of the story below (or on the foundation sill for the first story) and the bearing walls and partitions rest on the subfloor of each story.

Frieze: Any sculptured or ornamental band in a building; also the horizontal member of a cornice set vertically against the wall at its intersection with the soffit.

Frost line: The depth of frost penetration to soil.

Full round: Closet pole.

Fungicide: A chemical that is poisonous to fungi (e.g., chlorine bleach).

Furring: False work or framework necessary to bring the surface of a frame from where it exists to where it is desired to be.

Furrdown: To lower ceiling with false work for the purpose of hanging cabinets, lighting, etc.

G

Gable: The vertical triangular end of a building from the eaves to the apex of the roof.

Gable end: An end wall having a gable.

Gambrel: A symmetrical roof with two different pitches or slopes on each side (traditional barn); a ceiling in which the center is raised and the sides angle down to a junction with the top of the walls.

GFCI/GFI: A ground-fault circuit interrupter; a safety device that may be in the form of a specialized circuit breaker located in the panel but is more frequently found as an outlet; required in all exterior outlets, garage outlets, bathrooms, and all kitchen countertop outlets as well as any areas classified as damp or wet.

Girder: A timber used to support wall beams or joists.

Gloss enamel: A finishing material made of varnish and sufficient pigments to provide opacity and color but little or no pigment of low opacity; enamel forms a hard coating with maximum smoothness of surface and a high degree of gloss.

Gloss (paint or enamel): A paint or enamel that contains a relatively low proportion of pigment and dries to a sheen or luster; usually oil-based but now available in acrylic (water-based) enamels.

Grade, grading: As applied to finish drainage of residential construction, the establishment of a series of slopes and/or swales designed to remove the runoff water from the perimeter of the foundation or another undesirable location to a predetermined location; the minimum grade away from a home is a 5 percent slope or 6 inches of vertical fall within the first 10 feet of horizontal distance; any slope greater than 15 percent is not recommended.

Grain: The direction, size, arrangement, appearance, or quality of the fibers in wood.

Ground: An electrical term used to identify the green or bare conductor in a circuit as well as all conductors providing a path for a short circuit to an earth ground.

Grout: Mortar made so thin by the addition of water that it can be troweled into the joints and cavities of ceramic tile and fill them up solid; also, a material used to patch holes, cable ends, etc., in concrete.

Gutter: A shallow channel or conduit of metal set below and along the eaves of a house to catch and carry off rainwater from the roof.

H

Header: A structural member carrying the load of the wall above openings for windows and doors.

Headroom: The clear space between the floor line and ceiling, as in a stairway.

Hearth: The floor of a fireplace that extends into the room for safety purposes.

Heel of a rafter: The end or foot that rests on the wall plate.

Hip: The external angle formed by the meeting of two sloping sides of a roof.

Hip roof: A roof that slopes up toward the center from all sides, necessitating a hip rafter at each corner.

Honeycomb: Rough or void areas of concrete on the side of the slab caused by the stones having not been moved away from the form or air pockets that have not been removed by vibrating, jogging, or bumping the form.

HVAC (heating, ventilation, and air conditioning): The more accurate terminology to describe the entire scope of work done by AC contractors in residential construction; while these contractors are responsible for the heating and cooling equipment, they are also responsible for all venting of other items in the home, such as bath fans and dryers.

I

I-beam: A steel beam with a cross-section resembling the letter *I*.

I-joist: A manufactured wooden floor joist built of two laminated 2xes and a web of OSB.

Inspection plate: Coves that can be removed to check or service plumbing, usually only at bathtubs and showers.

Insulation, building: Any material high in resistance to heat transmission that when placed in the walls, ceilings, or floors of a structure will reduce the rate of heat flow; rated by R, or resistance value.

J

Jack or ram: Hydraulic devices used to apply tension to stressing tendons.

Jack rafter: A short rafter framing between the wall plate; a hip rafter.

Jack stud: Stud that is cut shorter than a precut stud and used to support areas under and above windows or doors and the sides of any opening to support headers; also called a *cripple stud*.

Jamb: The side piece of post of an opening; sometimes applied to the door frame.

Joint types:

- **Butt joints:** Squared ends or ends and edges adjoining each other.

- **Mortised joint:** A joint made by cutting a hole or mortise, in one piece and a tenon, or piece to fit in the hole, upon the other.
- **Scarfed joint:** A timber spliced by cutting various shapes of shoulders, or jogs, that fit each other.

Joint compound: A chemical mixture in powder or paste form for finishing joints, filling dimples, and covering nicks in gypsum wallboard.

Joint tape: A perforated paper strip that is embedded in joint compound to reduce the chance of cracking at joints in gypsum wallboard.

Joists: 2x lumber, I-joists, or truss joists designed to form the framing at the ceiling of a room to which ceiling materials are affixed; in two-story homes, they also serve as the floor joists above.

Joist hangers: Specially engineered metal hangers designed to allow attachment of joists flush with beams, etc.; they also eliminate the need for pressure blocking or toenailing.

K

Keeper/Strike plate: A metal plate that engages a mortise bolt or lock bolt when door is closed; attaches to the door jamb.

Kerf: A cut or slit made by a saw.

Kicker stakes: Stakes used at a 45-degree angle to form stakes that prevent the top of the form from leaning out under the pressure of the wet concrete.

Kiln drying: The process of drying sawn dimension lumber in a kiln or oven designed for that purpose.

Knee brace: A corner brace, fastened at an angle from wall stud to rafter, stiffening a wood or steel frame to prevent angular movement.

Knee wall: A short wall under a sloped ceiling, a service bar, or another wall shorter than the ceiling or partition height.

L

Landing: A platform between flights of stairs or at the termination of a flight of stairs.

Lattice: Crossed wood, iron plate, or bars.

Level: A term describing the position of a line or plane when parallel to the surface of still water; an instrument or tool used in testing for horizontal and vertical surfaces and determining differences of elevation.

Light: Space in a window sash for a single pane of glass; also, a pane of glass.

Linear foot: One running foot or 12 inches, measured in a straight line.

Lintel: A steel member usually formed in the shape of an angle iron used to support brick or other masonry units above; the piece of construction or finish, stone, wood, or metal that is over an opening.

Live end: The free end of a post-tension cable that carries the pocket former and cat head and projects through a hole in the form; also the end to which tension is applied in the reinforcing process.

Live load: The total variable weight on a structural member or system; includes the weight of people and furnishings; compare to a dead load.

Lookout: The horizontal return at the end of a rafter, or the construction that projects beyond the sides of a home to support the eaves and provide a flat surface on which the soffit is attached; also the projecting timbers at the gables that support the verge boards.

Louver: A kind of opening, generally in peaks of gables and the tops of towers, provided with horizontal slots that exclude rain and snow and allow ventilation; commonly decorative only with no ventilation function.

M

Makeup: The preparation of the foundation prior to placing concrete; includes digging of beams and installation of poly, post-tension cables, and necessary rebar.

Manual J: The AC plan calculation detail; that group of AC calculations that determines heat loads on a structure by analyzing its climate, location, orientation to the sun, mass, quantity of windows, and intended purpose; from these calculations are derived the tonnage and type of system required.

Masonry: Stone, brick, concrete, hollow tile, concrete block, gypsum block, or other similar building units or materials or a combination of the same, bonded together with mortar.

Mastic: A black, tarlike substance, also known as *cold roofing cement*, used on plumbing pipes in beams and where it projects through the slab as a sealing agent and bond breaker.

Member: A single piece in structure, complete in itself.

Metal lath: Sheet of metal that is split and drawn out to form an opening on which plaster or the bed for ceramic tile is spread.

Millwork: Generally, all building materials made of finished wood and manufactured in millwork plants and planning mills are included under this term; includes such items as inside and outside doors, window and door frames, wood shutters, porch work, mantels, paneling, stairways, moldings, and interior trim.

Miter: The joint formed by two abutting pieces meeting at an angle, usually 45 degrees.

Moisture content of wood: Weight of the water contained in the wood, usually expressed as a percentage of the weight of the oven-dry wood; also, a measurement taken of concrete slabs before laying some flooring materials such as wood.

Mold: Any of a class of fungi known to grow in the presence of moisture on cellulose-based material, such as wood, wood products, or paper.

Molding: A material, usually patterned strips, used to provide ornament variations of outline or contour, whether projections or cavities, such as cornices, bases, window and door jambs, and heads.

Molding types:

- **Lip:** A molding with a lip that overlaps the piece against which the back of the molding rests.
- **Picture:** A decorative molding shaped to form a support for picture hooks, often placed at some distance from the ceiling upon the wall to form the lower edge of a frieze.
- **Rake:** The cornice upon the gable edge of a pitch roof, the members of which are made to fit those of the molding of the horizontal eaves.

Mortise: The hole that is to receive a tenon or any hole cut into or through a piece by a chisel or mortiser, generally of rectangular shape.

Mullion: The construction between the opening of a window frame to accommodate two or more windows.

Muntin: The vertical member between two panels of the same piece of panel work; the vertical sash-bars separating panels of glass.

N

Newel: The principal post of the foot of the staircase; also, the central support of a winding flight of stairs; intermediate posts in a long run of stairs are also called *newels.*

Nonbearing wall: A wall supporting no load other than its own weight.

Nosing: The part of a stair tread that projects over the riser, or any projection; a term applied to the rounded edge of a board.

O

Ogee: A molding with a profile in the form of a letter S; having the outline of a reversed curve.

On center (OC): The measurement of spacing for studs, rafters, joists, and similar items in a building from the center of one member to the center of the next.

Oriented strand board (OSB): A manufactured wood panel formed through the random arrangement of large wood chips and exterior glue along with temperature and pressure; supposed to be dimensionally stable in all planes; used for roof sheathing and should never be left exposed for any length of time.

P

P-trap: A curved arrangement of piping usually under or adjacent to a plumbing fixture (e.g., sink, toilet, tub) designed to hold water and prevent the entry of noxious and possibly hazardous (e.g., methane) sewer gases into the structure.

Paint: Any of a number of opaque coatings used to weatherize or otherwise protect surfaces, both interior and exterior, and provide a pleasant, decorative appearance. May be oil- or water-based in flat, satin/eggshell, semigloss, or gloss finish, with the gloss finish being the hardest and most durable.

Panel: A large, thin board or sheet of lumber, plywood, or other material.

Partition: A wall that subdivides space within any story of a building.

Penny: As applied to nails, originally indicated by the price per hundred; the term now serves as a measure of nail length and is abbreviated by the letter *d*.

Piers: Masonry supports or pilings set independently of the main foundation; may be driven after the fact or bored with or without bell-bottoms as an integral part of the foundation pour.

Pigment: A powdered solid or liquid carried in suitable degree of subdivision for use in paint or enamel, which lends color and opacity.

Pilaster: A portion of a square column usually set within or against a wall.

Pile: A long post driven into the soil in locations in which it is difficult to secure a firm foundation; used many times for bridges and highways.

Pitch: Inclination or slope (also known as the *rake*), as for roofs or stairs, or the rise divided by the span.

Pitch board: A board sawn to the exact shape formed by the stair tread, riser, and slope of the stairs and used to layout the carriage and stringers.

Plan: A horizontal geometrical section of a building, showing the walls, doors, windows, stairs, chimneys, and columns used to cover the outside and inside wall surface.

Plaster: A pasty mixture of lime and gypsum used to cover outside and inside wall surfaces; predates the use of current gypsum board or drywall finishes.

Plasticity index (PI): An index that describes the soil's tendency to absorb water and, hence, expand and contract; A very sandy soil is at the low end, as it does not expand and contract when wet; clay, on the other hand, is at the high end.

Plat: Overall developer's plan showing home site dimensions, street locations, easements, and other physical features and descriptions of communities.

Plate: The top or bottom horizontal framing members of the walls of a frame building that rest on the foundation or other floor decks (base plate) and upon which the roof rests (top plate).

Plate cut: The cut in a rafter, which rests upon the plate; sometimes called the *seat cut*.

Plow: To cut a groove running in the same direction as the grain of the wood.

Plumb: Exactly perpendicular to level; vertical; true according to the plumb line.

Plumb cut: Any cut made in a vertical plane; the vertical cut at the top end of a rafter.

Plumbing grounds: The part of the plumbing system located within and beneath the foundation.

Plumbing top-out: Plumbing activity inside walls and attics in preparation for application of insulation and drywall.

Plumbing trim: Plumbing activity that includes the setting of all fixtures, connection, and testing to ensure proper operation.

Plumbing vent: Vertical pipes through the roof providing for the escape of noxious gases and odors and allowing air to enter the drainage system.

Ply: A term used to denote a layer or thickness of building or roofing paper as two-ply, three-ply, etc.

Plywood: A piece of wood made of three or more layers of veneer joined with glue and usually laid with the grain of adjoining plies at right angles; almost always an odd number of plies are used to secure balanced construction.

Pocket former/grommet: A cone-shaped device placed between the form and live-end anchor to form a pocket in the concrete at which stressors can hook onto the post-tension cable and a final clamp is attached before cutting the cables so that it can be grouted over flush with the slab surface.

Pop-off valve: More accurately termed the T&P, or *temperature and pressure, relief valve* on a water heater; designed to prevent catastrophic failures of water heater tanks due to excess pressure and/or temperature should there be a failure of the thermostat to control temperature; must be plumbed using a plumbing draw line and to the exterior of the structure.

Posttension cables: A system by which concrete slabs are reinforced with high-tensile-strength cables after the concrete has reached a hardness of approximately 2,000 psi.

Porch: An entrance way; usually covered.

Post: A timber set on end to support a wall, a girder, or another member of the structure.

Primer: The first coat of paint in a paint job that consists of one or more coats; establishes a base to allow for better coverage and adhesion of the top or finish coats; also, the paint used for the first coat.

Proctor density (PD): A soil density calculation based on a standard testing procedure developed by R.R. Proctor, a field engineer for the city of Los Angeles in the early 1930s; a given soil is collected and tested in the laboratory to determine its theoretical density, and the value derived therefrom is the 100 percent PD; the specifications in the field generally call for a PD of 95 percent for compacted soils; these tests are performed in the field by a testing lab and relate to the specified percentage of compaction as referenced back to the 100 percent value of previously collected soils.

Punchout: A term used in the construction industry to indicate the actions of a trade superintendent, builder, or construction manager during which the employee thoroughly double-checks their work and that of the others related for accuracy and completion.

Purlin: A horizontal framing member of the same dimension as the rafters positioned at a 90-degree angle to the rafters supporting several of them at one or more points; effectively shortens the span of rafters, allowing the use of a lesser dimension of lumber; any rafter longer than 12 feet should be supported by a purlin and purlin brace.

Purlin brace: A roof brace, not greater than 4 feet on center, which prevents the roof from sagging; ideally runs at a 90-degree angle from a rafter to a wall; there should be one Purlin brace for every other rafter longer than 12 feet, and it should be attached to both the rafter and the purlin.

Putty: A type of cement usually made into the consistency of dough and used for sealing glass in sashes, filling small holes and crevices in wood, and other similar purposes.

PVC: A plastic pipe made of polyvinyl chloride that is commonly used in drainage systems for plumbing; manufactured in various schedules or thicknesses (e.g., schedule 20 or 40) and in types intended for a specific use (e.g., CPVC is intended for hot water).

Q

Quarter round: A molding that presents a profile of a quarter-circle.

R

R-value: The resistance to the flow of heat through a combination of materials forming a wall, a ceiling, or another form of insulated construction.

Radiant heating: A method of heating that usually consists of coils or pipes placed in the floor, wall, or ceiling; not commonly used in the south.

Rafters:

- **Barge:** Non-weight-bearing rafter at the gable end nailed to lookouts.
- **Common:** Rafter that runs square with the plate and extends to the ridge.
- **Cripple:** Rafter that cuts between the valley and hip rafters.
- **Hip:** Rafter that extends from the outside angle of the plates toward the apex of the roof.
- **Jack:** Rafter that is square with the plate and intersects the hip rafter or valley to ridge.
- **Valley:** Rafter that extends from an inside angle of the plates toward the ridge or center line of the home.

Rail: The horizontal members of a balustrade or panel work.

Rake: The trim of a building extending in an oblique line, such as rake dado or molding; the inclination from the horizontal or perpendicular; applied generally to roofs.

Random-length lumber: Lumber delivered in assorted lengths, usually 12 feet or longer.

Rat run: 1-foot-by-4-foot or other light-framing members nailed to the tops of ceiling joists to prevent curling or rolling; mainly used with trusses.

Reglaze: The replacement of one or more broken panes of glass.

Reinforcing steel: Deformed steel rods (rebar) or metal fabric (mesh) placed in concrete slabs, beams, or columns to increase their strength.

Relative humidity: The amount of water vapor expressed as a percentage of the maximum quantity that could be present in the atmosphere at a given temperature; the actual amount of water vapor that can be held in a space increases with the temperature.

Return: The continuation of a molding or finish of any kind in a different direction.

Reveal: The depth of a window or door opening, usually measured from the surface of the exterior wall to the surface of the door or window; also, the marginal space, setback, or exposed face between one member and another, as in the reveal between the edge of a door jamb and the edge of the door casing; as a general rule, one expects all reveals of a given use to be equal throughout the home.

Ridge: The top edge or corner formed by the intersection of two roof surfaces.

Ridge board: The board placed on edge at the ridge of the roof to support the upper ends of the rafters; the top intersection of two adjoining pitched roof surfaces.

Rise: The vertical distance through which anything rises, as in the rise of a roof or stair.

Riser: The vertical board between two treads of a flight of stairs.

Romex: A woven-cloth type of wire sheath no longer in use; however, the term has continued to be used commonly to describe wiring used in residential construction.

Roof: The entire construction used to close in the top of a building.

Roofing: The material put on a roof to make it windproof and waterproof.

Roof jacks: Lead or plastic/rubber covers designed to fit over vent pipes protruding through the roof to prevent water leaks.

Roof sheathing: The boards or sheet material fastened to the roof rafters on which the shingle or other roof covering is laid.

Rowlock: Brick laid perpendicular to a brick wall, usually with a slight overhang or projection for appearance; sometimes sloped for water drainage, as in under windows.

Rubble: Roughly broken quarry stone.

Rubble masonry: Uncut stone, used for rough work, foundations, backing, and other similar functions.

Run: The length of the horizontal projection of a piece such as a rafter when in position.

S

S hooks: Wire hooks bent to maintain proper positioning of a posttension cable in a beam.

Sash: The framework that holds the glass in a window.

Saturated felt: A felt that is impregnated with tar or asphalt.

Scab: A short piece of lumber used to splice, or to prevent movement of two other pieces.

Scaffold or staging: A temporary structure or platform that enables workers to reach high places.

Scale: A short measurement used as a proportionate part of a larger dimension; the scale of a drawing is expressed as 0.25 inches = 1 foot.

Scarf: A joint between two pieces of wood that allows them to be spliced lengthwise.

Scratch coat: The first coat of plaster or tile mortar bed, which is scratched to form a bond for the second coat.

Screed: A long board or piece of metal used from form to form or level point to form that, when drawn across the surface of freshly poured concrete, creates a level surface and prepares for the bull float stage of finishing.

Scribing: The marking of a piece of wood to provide for the fitting of one of its surfaces to the irregular surface of another.

Sealer: A finishing material, either clear or pigmented, that is usually applied directly over uncoated wood for the purpose of sealing the surface.

Seat cut; plate cut: The cut at the bottom end of a rafter to allow it to fit upon the plat.

Seat of a rafter: The horizontal cut upon the bottom end of a rafter that rests upon the top of the plate.

Section: A drawing showing the type, arrangement, and proportions of the various parts of a structure; it is assumed that the structure is cut along a plane, and the section is the view gained by looking in one direction.

Setback: Required distance from the curb to the actual front of the home; the actual distance from the side property lines to the side of the home; the distance between a property line and the nearest point at which a building can be constructed, as established by local ordinance.

Sheath: As applied to posttension cables, the plastic sleeve or covering in which the tendon resides; it protects the cable from corrosion and prevents bonding of the cable to the concrete.

Sheathing: Wall boards and roofing boards that are applied under the weather or finished surface; forms the drainage plane of the structure.

Shellac: A transparent coating.

Shingles: A material that is laid in relatively small pieces that overlap each other to form an entire weather-resistant surface.

Shots and pins: A nail fired from a powder-actuated tool such as a Ramset, used to secure interior wall plates to the slab; used on 4-foot centers, within 6 inches of a doorway and at every butt joint in the plate.

Shutters: Originally light wood sections in the form of doors to close over windows to shut out light, give protection, or add temporary insulation; commonly used now for ornamental purposes only.

Siding: The outside finish (unless brick veneer) between the door and window casings and its associated trim.

Sills: The horizontal timbers of a house that either rest upon the masonry foundations or, in the absence of such, form the foundations.

Sleeper: A timber laid on the ground to support a floor joist.

Slump: The relative consistency or stiffness of a fresh concrete mix; the normal slump for residential construction should be 3 to 5 inches.

Soffit: The underside of the members of a building, such as staircases, cornices, beams, and arches, relatively minor in area as compared with ceilings.

Soffit vents: The portion of the attic ventilation system that is installed in the soffit of a structure to allow intake of cooler air than that which resides in the attic; in new home construction, it is common practice to install continuous soffit venting that extends down all sides of the home.

Soldier course: A course of brick in which they are laid so that they stand on end with the finished side face out.

Sole or sole plate: A horizontal member, usually 2 inches by 4 inches, on which wall and partition studs rest.

Span: The distance between the bearings of a timber, a piece of lumber, or an arch.

Specifications, specs: The written or printed directions regarding the details of a building or other construction; specs take precedence over plans.

Splash block: A small concrete block laid on the ground surface to receive roof drainage, prevent erosion, and carry it away from the building.

Splice: Joining or two similar members in a straight line.

Square foot: Equal to an area 1-foot-by-1-foot surface measure.

Square: A tool used by carpenters and others to obtain accuracy; a 90-degree angle; term applied to a surface 10 feet by 10 feet, or totaling 100 square feet.

Stair landing: A platform between flights of stairs or at the termination of a flight of stairs.

Stair rise: The vertical distance from the top of one stair tread to the top of the one next above; maximum allowable is 7¾" per the IRC.

Stairs, box: Those built between walls and usually with no support except the wall strings.

Starter shingles: First row of shingles with tabs removed and glue strip adjacent to the fascia with a 0.5-inch-by-1-inch overhang over the fascia; application in such a manner allows for the sealing of the bottom three tabs at the edge, which assists in wind resistance and seals gaps between the bottom tabs; some manufacturers now make a special shingle for this purpose.

Stool: The flat, narrow shelf forming the top member of the interior trim at the bottom of a window.

Stressing: The application of hydraulic forces through use of a ram to apply pull on posttension cables to achieve proper elongation, usually to 80 percent of the breaking strength.

String, stringer: A timber or another support for cross members; in stairs, the support on which the stair treads rest; also, stringboard.

Stringer: A long horizontal timber in a structure supporting a floor.

Strongback: Usually made of reusable form lumber; installed on top of ceiling joists, perpendicular to the joists; no reference to strongbacks as an acceptable bracing base exists in present code.

Stucco: A fine plaster used for interior decoration and fine work; also used for rough outside wall coverings.

Stud: An upright beam in the framework of a building.

Stud shoe: Metal reinforcement over a stud through which plumbing pipes pass.

Subfloor: A wooden floor that is laid over the floor joists and on which the finished floor is applied.

Swale: A shallow channel in the grading designed to allow water to drain off in the direction of the slope of the swale.

T

T-astragal: A T-shaped molding attached to one of a pair of swinging doors, against which the other door strikes; may also be referred to as an *astragal*.

T-poles: Temporary power poles installed by the electrical contractor and equipped with a meter by the power company to be used for power while the home is under construction.

Tail: The part of the rafter that extends past the exterior wall of the building.

Tendon: The complete assembly consisting of anchorages and post-stressing steel within the sheath; the tendons impart stress to the concrete after curing.

Termite shield: A shield, usually of noncorrosive metal, placed in or on a foundation wall or another mass of masonry or around pipes to prevent passage of termites.

Texture: Orange peel, spatter, rolled, sand, or another textured surface applied to drywall on wall per series specification.

Threshold: The beveled piece of wood, metal, stone, or concrete set at the sill of a hinged exterior door as a weather seal or to cover the seam of dissimilar flooring products.

Tie beam (collar beam): A beam situated in such a way that it ties the principal rafters of a roof together and prevents them from thrusting the plate out of line.

Timber: Lumber with a cross-section over 4 by 6 inches, such as posts, sills, and girders.

Toe kick: A recess under the front of a base cabinet at the floor line.

Toenail: To drive a nail at a slant with the initial surface to permit it to penetrate into a second member.

Topcoat: The exterior final coat of paint after the primer.

Top out: The in-wall installation (usually plumbing) behind the wall components of the mechanical, electrical, and plumbing systems of the home.

Transit level: A surveyor's instrument used to establish points and elevations; operates in both the horizontal and vertical planes.

Tread: The horizontal part of a step (minimum 10 inches in depth).

Trim: A term sometimes applied to outside of interior finished woodwork and the finish around openings.

Trimming: Putting the inside and outside finish and hardware upon a building.

Truss: Structural framework of triangular units for supporting loads over long spans.

Tub hangers: Metal clips nailed to the studs to support the back edge of a bathtub.

U

Undercoat: A coating applied prior to the finishing or topcoats of a paint job; may be the first of two or the second of three coats; in some usages of the word, it may become synonymous with *priming coat*.

Underground electrical service: Underground cable connection from the transformer or pedestal to the home; usually a direct bury cable not in a pipe; therefore, it is easily damaged.

Understory: Small brush or trees that are usually removed from a building site.

V

Valley flashing: Waterproofing the valley during the roofing process; this is done with either a metal pan of 28 ga or by using heavyweight roll-type roofing underlayment (typically 120-pound felt) below the shingles.

Valleys: The internal angel formed by the two slopes of a roof.

Vapor barrier: Material used to retard the flow of vapor or moisture into walls and foundations and thus prevent condensation within them (usually polyethylene).

Veneer: Thin sheets of wood.

Vent: A pipe installed to provide a flow of air to or from a drainage system or circulation of air within the system to protect trap seals from siphonage and back pressure.

Vestibule: An entrance to a house, also called a *foyer*; usually enclosed.

W

Wainscoting: Matched boarding or panel work covering the lower portion of a wall.

Wall, bearing: A wall that supports any vertical load in addition to its own weight.

Wall-out: Apply latex paint to all Sheetrock walls of a home.

Wane: Bark, or lack of wood or bark from any cause, on the edge or corner of a piece.

Warp: Any variation from straight in a piece of lumber.

Water cutoff: A valve usually found at two places on homes; one is at the water meter, and the other is located in the garage of attached-garage homes or the utility room of detached-garage homes.

Weather-strip: Narrow strips made of metal or another material designed such that when they are installed at doors or windows, they retard the passage of air, water, moisture, or dust around the door or window sash.

Weep holes: Spaces at every second mortar joint in the bottom course of bricks on a home in which the mortar is left out to allow for drainage of all moisture that collects on the drainage plane; also, allows for air circulation and drying of the interior surfaces behind the brick.

Window stool: The horizontal piece of finished wood lying flat on the framing on the interior side of the window; also known as the *windowsill*.

Wing wall: A decorative brick wall on the front of some homes, usually to give the home a larger appearance.

Wire nuts: Plastic cone-shaped devices in which a coil spring is arranged so that when they are twisted onto two wires to be connected, a solid connection is made.

Wood filler: A heavily pigmented preparation used for filling and leveling off the pores in open-pored woods.

XYZ

Zero lot line: A term referring to the manner in which a home is situated with no setback from one property line; **important note:** No openings may exist on the zero-lot-line side, including doors and windows and soffit, dryer, or other vents.

About the Contributors

Kerry R. Gilbert is the former president of Kerry R. Gilbert & Associates, Inc., a land planning and development consulting firm located in Houston. The firm specialized in residential and commercial subdivision planning, major thoroughfare planning, master planning, government agency review and approval process, state and local ordinances regulating development, and analysis and potential land use studies for acreage tracts. Kerry R. Gilbert & Associates, Inc., was widely recognized for its innovative design concepts and has been the recipient of numerous national awards presented by the National Association of Homebuilders, the Pacific Coast Builders Conference, and *Professional Builder* magazine.

Prior to forming Kerry R. Gilbert & Associates, Inc., in 1985, Mr. Gilbert was assistant director of the City of Houston's Department of Planning and Development, where he was in charge of all development-related activities. Mr. Gilbert was with the Department of Planning and Development for 13 years and instrumental in the drafting of the city of Houston's Development Ordinance, which is the basic document regulating development in the city of Houston and its extraterritorial jurisdiction. For his efforts, Mr. Gilbert was named Outstanding Contributor to Planning in 1984 for the state of Texas by the American Planning Association.

In 1997, Kerry R. Gilbert & Associates, Inc., partnered with DeForest Design Group, Inc., to form KGA/DeForest Design, LLC. KGA/DeForest Design, LLC, provides landscape architectural design and consultation services for residential communities, commercial developments, and numerous public institutions including cities, counties, and municipal utility districts. See the company's website (www.kgadeforest.com) for additional information and projects. Mr. Gilbert remains an owner and a board member of the company today.

Mr. Gilbert currently serves on the board of directors for the Katy Area Economic Development Council as well as the Fort Bend County Development Advisory Committee and the board of directors of the Texas Heritage Parkway Improvement District.

Mr. Gilbert is currently in semiretirement as he serves as president of KGA Consulting, LLC, a land planning and development consulting firm that provides consultation services to select clientele.

Prior to retirement, Mr. Gilbert's professional affiliations included:

- American Planning Association
- Greater Houston Builders Association
- Fort Bend Chamber of Commerce
- Katy Area Economic Development Council
- Katy Area Chamber of Commerce
- Urban Land Institute
- West Houston Association board of directors (ex-officio)

Throughout his career, Mr. Gilbert served as a lecturer and panelist for the following organizations:

- Urban Land Institute
- American Planning Association
- Texas Association of Builders
- Texas Surveyors Association
- West Houston Association
- Houston Bar Association—Real Estate Section
- Houston Board of Realtors
- Houston Community College
- University of Houston

Michelle Stoops is a native Houstonian and a graduate of the University of Houston. *Go, Coogs!*

During college, Ms. Stoops worked for her dad's land planning consulting firm, Kerry R. Gilbert & Associates, Inc., and while being mentored by her dad and learning all aspects of a land planning consultant, she stayed with the company for another 25 years.

Now, Ms. Stoops is the owner of KM Stoops Services, Inc., a company that offers a variety of services to developers and builders, including graphics, commercial and residential guidelines, and architectural review services.

Figure Credits

IMG A.1.1: Copyright © by Country Park Portraits. Reprinted with permission.

IMG A.1.2: Copyright © by LC Photographers. Reprinted with permission.

References

HOMEBUILDER CONSTRUCTION MANUALS

McGuyer Homebuilders, Inc. 2003. *Construction and Customer Service Manual.*
Village Builders. 1995. *Construction Standards.*
General Homes Corporation. 1984. *Construction and Customer Service Manual.*
Lennar Homes. 2001. *Construction Manual.*
David Weekley Homes. 1994. *Construction Manual.*
Builder Unknown. ca. 1950s. *Homebuilding Construction Manual.*
D.R. Horton Homes. n.d. *Construction Manual.*
General Homes Corporation. Ca. 1970s or 1980s. *Take-Off Handbook.*
Royce Homes. 1999. *Construction Manual.*
Trendmaker Homes. 1997. *Construction Standards.*

BOOKS

Moore, Francis C. 1897. *How to Build a Home.*
Lester Suttcliffe, L. 1899. *Modern House Construction*, Vols. I, II, III, IV.
Helm, W.H. 1921. *Homes of the Past.*
Mercer, Henry C. 1924. *Origins of Log Houses in the United States.*
Bemis, Albert Farwell, and John Burchard. 1933. *The Evolving House.*
Johnstone, Kenneth. 1950. *The Business of Homebuilding.*
Pickering, Ernest. 1951. *The Homes of America.*
Housing and Home Finance Agency. 1953. *Construction Financing for the Home Builder.*
Lucas, Edgar. 1957. *Modern House Building*, Vols. I, II, III.
Chinitz, Benjamin. 1964. *City and Suburb: The Economics of Metropolitan Growth.*
Goodkin, Lewis M. 1974. *When Real Estate and Home Building Became Big Business.*
Weiss, Marc A. 1987. *The Rise of the Community Builders.*
Rourke, Dennis J. 1994. *The American Home Builder and the Housing Industry.*
Jacobus, Charles J. 2000. *Texas Real Estate Law.*
Woodson, R. Dodge. 2000. *Be a Residential Land Developer.*
Johnson, David E. 2002. *Residential Land Development Practices.*
Benton, W. C., and Linda F. McHenry. 2010. *Construction Purchasing and Supply Chain Management.*
Allen, D., R. Thallon, and A. Schreyer. 2017. *The Fundamentals of Residential Construction.*
Jensen, Thomas L. 2017. *Texas Real Estate Principles 1.*
Stevens, Robert. n.d. *Home Building A to Z.*

MAGAZINES AND JOURNALS

Scientific American, Architects and Builders Edition, Munn & Company, August 1891.
Building Age, Architectural and Building Press, Inc., several editions, 1920s.
American Builder and Builder Age, Simmons-Boardman, several editions, 1930 and 1932.
Journal of Home Building, National Association of Homebuilders, September 1957.

COMMERCIAL AND ACADEMIC ONLINE RESEARCH RESOURCES

The Texas Real Estate Research Center. wee.recenter.tamu.edu.
Builder magazine, 2022 Builder 100.
Metrostudy reports via business journals (Houston, Austin, San Antonio, and Dallas).
Frame. frame.builderonline.com/markets/all Texas markets.